RYAN PARRY

Quantum Watchers: The Nexus Paradox

I

Part One

1

The Watchers

The night sky glimmered with stars as Dr. Eva Gray gazed up at the cosmos through her telescope. As an astrophysicist, she had spent countless nights analysing the mysteries of the universe. But tonight felt different. There was an electric charge in the air, a sense of anticipation.

Eva's radio telescope was aimed at a recently discovered exoplanet in the Andromeda galaxy. Dubbed Pandora-3, it orbited a red dwarf star over 2 million light years away. Pandora-3 exhibited strange quantum signatures that defied explanation. Eva's fellow researchers dismissed the anomalies as instrument errors, but her instincts told her there was more to this enigmatic world.

She fine-tuned the telescope's frequency, honing in on the tantalizing energy signals. What she saw made her pulse quicken. A complex sequence of quantum flux patterns - almost like a message.

"It can't be..." she whispered. Was this evidence of an intelligent civilization? One that understood advanced quantum mechanics? The revelation left her stunned. She urgently documented her observations, knowing this changed everything about how humans understood their place in the cosmos.

Eva sent an encrypted data packet to her colleague, Dr. Vikram Singh, a quantum physicist at Caltech. Moments later, her computer buzzed with an incoming video call.

"What am I looking at Eva?" Vikram asked, studying the data with bewilderment.

"This is first contact," Eva said quietly. "We always assumed we were alone, but we were wrong."

"I've never seen quantum wave functions behave this way naturally," Vikram said. "There must be an advanced civilization manipulating quark spin on a massive scale. But how is this possible?"

"We have to gather more data. I'm going to book time at the LIGO observatory to analyse Pandora-3's gravitational waves. That will give us a fuller picture."

"Be careful, Eva. We're entering uncharted waters," Vikram warned. "Consider what happened the last time humanity made first contact."

He was referring to the disastrous encounter with the Tau Ceti civilization decades ago. The Tau Ceti had manipulated quantum probabilistic wormholes to travel to Earth but panicked after being met with human hostility. Their botched wormhole escape destroyed Pluto and sparked an intergalactic war.

"This is different," Eva countered. "The Watchers—I think that's what we should call them—they're benevolent. They knew to contact us gently using quantum mechanics, the universal language of science. We need to reply."

Vikram weighed her words and nodded. "Very well, but we should proceed with caution. I'll speak to the Planetary Defence Council and try to secure us some time on LIGO. Just promise me you won't do anything rash."

"I promise," Eva said. After Vikram signed off, Eva gazed once more through her telescope at Pandora-3. She would soon make history by establishing humanity's first peaceful exchange with alien intelligence. A thought occurred to her… the Watchers knew so much about quantum manipulation. Perhaps they held the key to illuminating the most confounding quantum mysteries: wave-particle duality, entanglement, measurement, and observation.

This was humankind's opportunity to advance its science and technology for millennia. Eva felt the weight of responsibility on her shoulders. If she hadn't discovered that anomalous signal, the world might never have known what existed in the farthest reaches of space. She silently thanked the Watchers for reaching out across the cosmos and for taking the first step toward intergalactic friendship.

2

The Quantum Anomaly

Eva arrived at the LIGO observatory in Washington, eager to analyse the gravitational waves from Pandora-3. She met Vikram in the control room, which buzzed with researchers reviewing data monitors.

"Thank you for getting us access on such short notice," Eva said. "With these gravity wave readings, we may finally understand how the Watchers are manipulating quantum phenomena on their planet."

"I briefed the director of LIGO on our discovery," Vikram replied. "While anomalous, he agrees the implications could be enormous."

An astrophysics coordinator led them into a lab with screens displaying colourful frequency charts. He explained that the preliminary gravitational wave data from Pandora-3 showed highly abnormal oscillations.

"We double- and triple-checked our instruments. The readings are accurate if mystifying," he shrugged.

Eva and Vikram pored over the data in stunned silence. Pandora-3's gravitational waves pulsed in ways that defied the laws of physics.

"These patterns couldn't occur naturally," Eva whispered. "The Watchers must be engineering the gravitational field around their planet, but how?"

"This is beyond any known science," Vikram said. "To influence gravity waves on a planetary scale would require energy surpassing any power source humans have conceived."

Their analysis was interrupted by an urgent alert. Across the observatory, screens flashed with warnings as stunned researchers looked on.

"What's happening?" Eva asked the coordinator.

"Our satellites detected massive spikes of X-rays and gamma radiation emanating from the Andromeda Galaxy," he said worriedly.

Eva and Vikram exchanged alarmed looks. They rushed to the satellite monitoring station, where screens displayed maps of the Andromeda Galaxy highlighted with flaring heat signatures.

"The cosmic radiation is being emitted from the region containing Pandora-3," a lead researcher said gravely.

Eva noticed the radiation levels increasing at an exponential rate. "At this rate, Pandora-3 will be bathed in lethal radiation within hours!"

"But what's causing this?" Vikram asked. "There must be forces at play here we can't yet comprehend."

Eva's mind raced. She recalled a branch of quantum mechanics called vacuum energy—the theory that empty space is not blank but seethes with virtual particles and antiparticles winking in and out of existence. Those ephemeral quantum units contained immense latent energy.

Somehow, the Watchers were tapping into vacuum energy to fuel their unfathomable quantum technologies. But their experiments had also triggered a runaway reaction, resulting in the deadly radiation engulfing their planet.

"We have to help them!" Eva said. "The Watchers don't realize the catastrophe their experiments have unleashed."

Vikram's face was grim. "If the radiation reaches critical levels, the entire Andromeda galaxy could become unstable. Pandora-3 may even implode into a black hole."

As scientists across the observatory scrambled to model catastrophe scenarios, Eva noticed new data from the gravitational wave detectors. The patterns had become wildly erratic, suggesting Pandora-3's very existence was threatened.

She rushed to transmit a message toward Andromeda using the observatory's powerful SETI antenna array. "Watchers, this is humanity. Your experiments have created a quantum anomaly, putting your civilization in mortal danger.

For hours, she sent warnings and coordinates for shielded areas on Pandora-3, hoping the Watchers might evacuate to safety. But she knew that if the runaway reaction continued unchecked, even her desperate communications might not be enough to avert disaster. The first intelligent species humanity has ever encountered could be on the brink of annihilation.

3

Entering the Nexus

Eva paced the control room as researchers at LIGO continued to monitor the dire situation on Pandora-3. The strange radiation emanating from the planet was increasing, suggesting its quantum experiments were becoming more unstable.

"Have the Watchers responded to any of our messages?" she asked Vikram anxiously.

He shook his head. "I'm afraid not. If they did evacuate to shielded areas, they may have lost the ability to communicate."

Eva frowned, trying to think of alternatives. "What if we could send a ship to Andromeda through some kind of wormhole? We might be able to provide assistance on the planet directly."

Vikram gave her a patient look. "Intergalactic travel via wormholes is purely theoretical right now. Our civilization simply doesn't have the technological means."

Their conversation was interrupted by an alert. Across LIGO, alarms blared as screens flashed warnings about an approaching wave of cosmic energy. Researchers braced themselves at their workstations.

"Analysis indicates the source is Pandora-3," the coordinator announced grimly.

Eva watched the screens with growing unease. "This is no ordinary radiation. I think we're detecting the first test pulses of an experimental quantum energy beam."

"You believe the Watchers are attempting to transmit to Earth?" Vikram asked.

"It's possible. The radiation levels may be too severe for normal communication. This beam could be a last-ditch effort to reach us."

As she studied the incoming energy waveforms, an idea took shape in Eva's mind. The beams were directed at LIGO with surprising precision. What if the Watchers were manipulating quantum entanglement in order to pull Earth into a transitory wormhole?

"We need to send a response transmission," she told Vikram excitedly. "If we can establish an entangled link with their beam, I believe we can create a temporary nexus for information exchange. Maybe even transportation."

Vikram looked uncertain. "That would be an immense feat of quantum engineering. Are you certain humanity is ready for that?"

"We have to try," Eva insisted. "This may be the Watchers' only chance for survival."

After a tense discussion, LIGO's director approved Eva's plan. Researchers hurriedly calibrated the antenna array to emit a precisely entangled return beam toward Pandora-3.

As the observatory's dish telescopes hummed to life, Eva initiated the transmission. Dizziness washed over her as she felt an intense tug on her consciousness from the streaming quantum energy. Gasps from the control room told her others felt it too.

Blinding light enveloped the facility. When it receded, Eva found herself standing in an otherworldly alien chamber. Vikram and the rest of the team appeared equally disoriented as they took in their surreal surroundings.

Lights flickered, and containers holding strange fluid samples lined gleaming shelves. A doorway led to a vista of swirling pink and purple nebulas streaking an indigo sky.

"The nexus worked," Eva breathed. "We're inside a quantum research facility on Pandora-3."

A tall figure entered the chamber and froze at the sight of Eva and her colleagues. Its willowy frame, large, luminous eyes, and shimmering skin marked it unmistakably as one of the Watchers.

Eva stepped forward slowly. "Do not be frightened. We received your signal and want to help."

The Watcher appraised her silently, then spoke in a musical trilling language. Realizing it was telepathic, Eva opened her mind to listen. To her excitement, she found she could understand the alien's thoughts.

"The experiment has become unstable. Your warning message was received too late," the Watcher conveyed. "But we have now established a quantum entanglement bridge with your planet. Together, our civilizations may survive."

Eva exchanged relieved looks with Vikram. The first contact had been successful. Now they needed to work quickly to avert disaster before the deadly energy cascade engulfed this world and their own.

II

Part Two

4

Paradox Lost

Eva followed the Watcher down metallic corridors, exiting the quantum laboratory onto an observation platform. Her mouth fell open at the sight.

Below stretched a vast alien metropolis unlike anything she had ever seen. Sleek towers spiralled into the sky, connected by suspended walkways humming with transport pods. Lush gardens and fountains created vibrant oases between buildings.

The city shimmered with a preadolescent glow, as though forged from the same luminous material as the Watchers themselves. Eva guessed this bioluminescence resulted from the Watchers' ability to manipulate vacuum energy, converting it into light and fuel.

Her amazement turned to dismay as she noticed plumes of smoke rising from smashed structures. Fires raged in several sectors. The city had already suffered severe damage from the quantum accident.

"How long do we have until the cascade reaction reaches your capital?" Vikram asked the Watcher urgently.

"We estimate cycles," it replied telepathically, a note of despair entering its thoughts. "Our annihilation is imminent."

Eva placed a comforting hand on the alien's slender arm. "We will stop this together. Your civilization created extraordinary quantum technologies. There must be a way to contain the reaction."

The Watcher considered this, head tilted. "Follow me," it finally conveyed.

"There is one possibility, but it is perilously risky."

They descended through levels bustling with more Watchers. Although their expressions remained inscrutable, Eva sensed the anxiety lying beneath their still surfaces.

At the lowest tier, the Watcher led them into a vault secured by shimmering force fields. In the centre stood an intricate machine ringed with crystal spikes.

"A quantum time portal," Vikram gasped. "But that's only theoretical…"

"It is functional but uncontrollable," the Watcher responded. "We did not foresee its power to tear holes in the chronal fabric of our universe. However, it may be our only means of sealing the quantum anomaly."

Eva's mind raced. By opening a controlled wormhole to the past, they could prevent this version of Pandora-3 from ever conducting the calamitous vacuum energy experiment. But the Watcher was right—the risks were profound.

She turned to Vikram. "If we work together to stabilize the spacetime curvature, I believe we can direct the portal to before the accident occurred, avoiding this entire chain of events."

He hesitated, then nodded resolutely. The two of them joined efforts with the Watchers to calibrate the temporal vortex. As they worked, seismic quakes rocked the planet's surface. Fires erupted in neighbouring sectors. They were running out of time.

"Initiating gateway sequence," the Watcher announced.

The portal crackled and swirled like a crystal cyclone. Eva could see faint glimpses of the past through the shimmering chronal field. With careful manipulation, she nudged the vortex's spacetime coordinates to the moment they needed.

"Now!" she yelled.

The Watcher activated the portal. Blinding light flooded Eva's senses. When her vision cleared, she found herself standing beside Vikram in the LIGO control room. Dazed, she realized they had returned to the moment before receiving the Watchers' call for help.

"Did…did it work?" Vikram asked hesitantly.

Eva rushed to the satellite monitors. To her relief, they detected no abnormal activity in the Andromeda Galaxy. Pandora-3 was calm and intact, unaware of the disaster it had narrowly avoided.

By leaping into the past, Eva and Vikram had severed this branch of the causal timeline before its terrible outcome. The paradox had been eliminated.

Eva knew the Watchers were still out there, existing without knowledge of what had transpired. But someday, when their civilization was ready, humanity would make contact again and forge a powerful alliance across the stars.

5

Colliding Worlds

Life returned to normal for Eva and Vikram in the weeks after their extraordinary experience on Pandora-3. With no anomaly to investigate, the encounter felt surreal, like a strange dream.

Eva focused her research on new mysteries: mapping dark matter and seeking exoplanets with biosignatures. But on quiet nights at the observatory, her mind often wandered back to the Watchers. Were they still deciphering the secrets of quantum mechanics, unaware of the catastrophe they had narrowly avoided? Would humanity one day reunite with its alien friends?

Late one night, Eva sat scanning extragalactic signals when an urgent alert flashed on her monitor. She quickly called Vikram.

"Are you seeing this energy spike?" she asked. "It's off the charts."

"I'm looking at it now," he replied. "The epicentre appears to be near Pandora-3 again."

Eva's heartbeat quickened. Were the Watchers in peril once more? She rushed to access LIGO's satellites to better analyse the disturbance.

As she studied the data on her screen, she realized with a start that this was no quantum accident. The energy signals were too precise, propelled by an obvious intelligence.

"These waveforms have a message embedded in them," she told Vikram excitedly. "I'm going to run it through our translation algorithm."

Moments later, a sequence of panicked alien symbols appeared on her display. Eva's gut clenched as she deciphered their meaning.

"It's a distress call... from a civilization unknown to the Watchers," she said. "Their planet is experiencing severe gravitational disturbances, pulling it into an uncontrolled collision course with Pandora-3!"

"Two advanced civilizations, about to crash into each other?" Vikram said this in alarm. "The impact would be apocalyptic."

Eva's mind raced. According to the message, the alien society possessed technology, allowing them to manipulate wormholes. Perhaps the same technology that had torn a rift in time on Pandora-3...

"I have an idea," she told Vikram. "We need to send an urgent warning to the Watchers about this impending collision, and instructions for generating a stable traversable wormhole."

Eva swiftly transmitted her message to Pandora-3 in the cosmic lingua francs of mathematics and physics. She anxiously waited, hoping the Watchers possessed the knowledge to implement her suggestions.

Hours later, LIGO detected increased energy emissions from the Andromeda galaxy. Eva saw with relief that they were consistent with an opening wormhole.

"It's working!" Vikram exclaimed. "The gravitational waves indicate the alien planet is getting pulled through the portal."

Moments later, instruments confirmed that the unknown planet had vanished from its collision path. The wormhole had transported it safely to another galaxy, averting disaster.

Eva allowed herself an exhale of relief. Thanks to the Watchers' swift response, two civilizations had been saved, and humanity had once again assisted its alien allies.

In the weeks that followed, Eva became consumed with deciphering the distress message's alien language. It clearly belonged to a highly sophisticated species. She wondered...could they have developed the time portal technology that had caused so much havoc on Pandora-3?

If so, humanity would need to exercise caution in any future encounters. One thing was certain - with multiple advanced civilizations emerging across the cosmos, the universe was becoming smaller and more intertwined. The age of isolation was over. The time of connection had begun.

One night, as Eva gazed up at the Andromeda galaxy's faint glow, she sensed the Watchers also looked to Earth, awaiting the day their two peoples could unite to explore the wonders of the universe. She smiled, knowing that an incredible future was within humankind's grasp.

6

Crossing Timelines

The years passed in a blur as Eva's work took her across the globe, advancing humanity's understanding of the cosmos. She led quantum gravity experiments at CERN, unlocked hidden secrets in gravitational wave data, and pushed the boundaries of astrophysics.

Through it all, she held onto the hope that she would one day reconnect with the Watchers. Their profound insights into quantum entanglement, vacuum energy, and wormholes continued to inspire Eva's research. She knew an alliance with the aliens could fast-track humanity's technology by eons.

So when a signal bearing the Watchers' signature was detected, Eva immediately departed for the remote radio observatory processing the transmission. As her self-piloting shuttle broke through the atmosphere, she reviewed the cryptic message with growing unease.

Rather than a greeting, it was a series of emphatic warnings about the hazards of "chronosynchronicity" and "inter-stream transits." The Watchers claimed any further contact would risk another devastating time anomaly.

Eva was baffled. After years of silence, why were they suddenly discouraging communication? Had she and Vikram's wormhole experiment truly been so dangerous?

Her shuttle landed smoothly at the sleek observatory complex, forged from eco-friendly metals and graphene. Eva was greeted by Vikram and the lead researchers, who looked relieved to see her.

"I assume you decoded the message," Vikram said. "Any theories on why our friends are trying to ghost us?"

"I have no idea, but I say we reply anyway," Eva answered. "We need to show them we've matured as a civilization."

The team spent hours fine-tuning the powerful narrow-beam transmitter to send a message back to Pandora-3. In her communication, Eva provided updates on humanity's scientific progress as well as thoughtful arguments for maintaining their cosmic connection.

Days passed with no response as the researchers anxiously monitored sensors directed at the Andromeda Galaxy. Finally, faint return signals were detected emanating from Pandora-3.

As the message decrypted on their screens, Vikram looked puzzled. "They're giving us complex directions for... constructing an interstellar quantum communicator? Why not just use our existing channels?"

Eva's brow furrowed. "Unless...they can no longer receive our conventional transmissions in their spacetime realm."

A daring hypothesis took shape in her mind. Their previous wormhole journey must have created a second alternate timeline - one where Pandora-3 continued its perilous vacuum energy experiments unimpeded. Somehow, their message beams were crossing between parallel quantum realities.

Eva proposed modifying the observatory's dish array to transmit along the higher-dimensional wavelength the Watchers specified. With some reluctance, Vikram and the lead engineers agreed.

They spent several days aligning the telescopes to pierce through divergent timelines. At last, the quantum beacon activated, flooding space with intricately entangled photons.

Moments later, the observatory filled with shimmering light as a portal opened, much like the one on Pandora-3 years earlier. Only this time, Eva recognized the Watcher who stepped through.

"You helped us avert disaster once before," it spoke telepathically. "But in this timeline, your warning came too late. Our unchecked experiments have made Pandora-3 unstable. We require your aid again."

Eva clasped the alien's slender hand. "Of course, old friend. Though our worlds now exist on separate plains, humanity stands ready."

Fascinated by the implications of parallel time streams, Eva joined the Watcher in preparing a chronic-shift vessel for Vikram and herself to visit Pandora-3. If they could replicate their past success, both realities would be preserved.

As the ship launched into the swirling nexus between planetary timelines, Eva marvelled at the new frontiers of exploration opened by quantum mechanics. Guarding the integrity of spacetime and building bridges between worlds would be her life's work. Wherever the universal wave function led, humankind would boldly follow.

7

Temporal Disturbances

The chronic-shift vessel emerged from the shimmering nexus into the familiar sight of Pandora-3's settlements. But as Eva gazed out the viewport, her stomach clenched. Rather than gleaming towers, she saw only crumbling ruins.

This was a world on the brink of collapse. Without the warning from Earth's timeline, the vacuum energy accident had escalated unchecked. Pandora-3 now resonated with dangerous quantum temporal distortions.

Eva exchanged a worried look with Vikram. Preventing disasters would be harder this time.

They docked beneath a flickering energy dome, maintaining an atmosphere in one of the last intact sectors. The waiting Watcher led them hurriedly through the compound.

"We calculate temporal disintegration within cycles," it warned. "Multiple contradictory time streams are splitting off from this version of events, straining cosmic coherence."

"Temporal tuning forks," Vikram suggested. "They could help anchor your prime timeline."

The Watcher nodded and showed them a chamber where technicians fine-tuned towering bronze constructs resonating at precise quantum frequencies. Eva recognized them as tools to keep timelines in harmony—one of many astonishing Watchers technologies.

A tremor shook the sector, sending equipment crashing as Eva and the others struggled to keep their balance. Alarms blared, signalling another localized time fracture.

Eva studied the readings flashing on screens. "At this fracture rate, your world will be fragmenting faster than you can repair it. Is there any way to identify the root cause?"

The Watcher brought up holographic models of Pandora-3's quantum strata. Gesturing to the layers, it explained, "Prior to the vacuum disaster, our civilization harnessed quantum temporal resonance for energy. But the accident warped our planet's chronal field."

Studying the distorted quantum matrix, Eva noticed timeline branches clustering around one specific set of coordinates on the planet's surface.

"There - a temporal nexus," she realized aloud. "Your people tapped into vacuum energy there, not realizing it would become a rupture point in spacetime."

"Are you able to seal the nexus?" the Watcher asked hopefully.

Eva hesitated. Quantum chronology was far beyond human knowledge. But she had an obligation to try.

She discussed ideas with Vikram and the Watchers to construct a quantum temporal shield—a device that could cloak the vulnerable nexus point from turbulent cosmic time stream forces. It was an outrageous undertaking, requiring intricately calibrated graviton mesh weaving and tachyon arrays. But if successful, it could mend this fractured timeline.

Continuing quakes and distant explosions resonated through the compound as they worked feverishly to assemble the shield components. Vikram and Eva provided engineering expertise, while Watchers manipulated technologies unfathomable to humans.

Barely an hour later, the unsettled sky began swirling with interdimensional eddies—a sign of impending chronological collapse. It was now or never.

Led by the Watchers, their team transported the shield apparatus to the nexus coordinates outside the crumbling capital. They braced themselves against violent localized time distortions assaulting the area.

"Activating quantum cloak in 15 micro spans," the lead Watcher signalled.

"Cross-temporal harmonization must be precise."

Eva closed her eyes as the shield powered up, sending out brilliant tachyon rays to envelop the nexus point. For agonizing seconds, chronal forces warped chaotically around them. Then finally, she felt space-time assume a gentle harmonic balance.

Opening her eyes, Eva saw that the sky had stabilized. The nexus had been sealed out of phase from the time stream. Pandora-3 was whole once more.

She turned to the Watcher with a weary but grateful smile. This parallel universe now has hope for a future.

8

Reality Unravelling

In the months following the narrowly averted planetary collapse, Eva stayed in close contact with the Watchers. She volunteered her knowledge to help rebuild their damaged settlements and technologies.

The quantum temporal shield had saved Pandora-3, but many mysteries remained. The accidents unleashing dangerous exotic energy continued to elude the Watchers' understanding.

Eva theorized that their early experiments trying to harness vacuum energy had permanently destabilized the local spacetime curvature. Like a pulled thread unravelling the cosmic fabric, each new effort to manipulate quantum phenomena further unravels reality.

She discussed her concerns in an urgent meeting with the Watchers' scientific council. "Your innovations in quantum engineering are astonishing," she told them. "But tapping vacuum energy clearly endangers the integrity of space-time. Perhaps this research should stop."

The lead Watcher linked minds with the others before responding. "Abandoning such work would surrender immense potential. However, your warning warrants reflection. We shall re-examine the ethical implications."

Eva left the discussion feeling hopeful. She knew the Watchers valued harmony and knowledge above all else. If her advice resonated with even a few, perhaps they would steer their civilization toward safer frontiers.

Her optimism soon faded. Scanning Watchers' communications in the following weeks, she saw no announcements restricting quantum energy experimentation. If anything, their efforts grew bolder and more frequent. The accidents were simply considered engineering challenges to solve, not reasons to pause.

When a mishap during a vacuum trial nearly caused mass gravitational inversion, Eva decided she could no longer stay silent. She confronted the lead Watcher, pleading with him to convince the government to halt research endangering reality itself.

The alien's expression remained impassive, but his mental voice conveyed profound sadness. "I confess you speak wisdom, dear friend," he shared. "But our people will not abandon the potential rewards, despite the risks. We have become blinded by pride and idealism."

He went on to make a dire confession. For generations, Watchers scientists had detected their universe slowly deteriorating—stars dying faster, galactic clusters falling out of alignment. Vacuum energy was the only force promising to reverse the decay. They could not relinquish it now.

Eva reeled at this revelation. She realized now the true scope of their civilization's desperation. Even if she convinced them to stop this research, they would never abandon their doomed cosmos without hope of a solution.

Unless...she could give them hope, After months of collaborating with the Watchers, Eva finally grasped the intricate commune their people shared with nature, time, and cosmic cycles. There was a way forward if they could see it.

"Entropy and decay are built into our universe's structure," she explained. "Clinging too tightly to permanence will only create suffering. But accepting impermanence opens new possibilities!"

She shared human philosophies that found solace on the paths of mindfulness, compassion, and renewal. And she outlined new energy models based on decentralized networks and renewable influxes from stars.

The Watcher was silent for a long moment after her entreaty. Finally, he took her hands gently. "Your perspective holds wisdom beyond your years and beyond our own. We will share this message."

In the weeks that followed, Eva was invited to speak before assemblies of

Watchers. She counselled embracing change while approaching research and technology with care. Many took her lessons to heart, and the tide slowly began to turn.

Nearly a year later, the Watchers' governing quantum council unanimously halted vacuum energy testing. They announced a new era for their civilization—one embracing sustainability, balance, and cosmic harmony over manipulation. A choice that signalled hope for the future.

9

The Fabric of Space-Time

Eva stepped through the portal connecting Earth to Pandora-3, emerging into a scene of orderly industriousness. Watchers glided efficiently about the station, monitoring instruments and calibrating the wormhole nexus.

Since the decision to halt risky vacuum energy research, the Watchers have found their footing again. Focusing their technology inward for knowledge and sustainability nourished their minds and planet.

Eva was proud of the small role she played in guiding their civilization. And she was thrilled they now invited select humans to study with them.

Her long-time colleague, Vikram emerged from the portal beside her. His face lit up, taking in the alien research site. "Can you believe we're standing in another galaxy? This alliance with the Watchers could propel humanity centuries forward!"

They were led down curving passages to meet with the project director. The willowy Watcher greeted them warmly. "We are honoured by your presence. You will hold the first human seats in our recently convened Parliament of Worlds."

Eva's eyes went wide. "Does this mean other civilizations have joined your Federation?"

"Indeed. Eight spacefaring cultures from the Andromeda Council worlds have come together. We are only the beginning."

Eva shook her head in wonder. Just months earlier, the Watchers stood on the brink of collapse. Now they were leading a pioneering effort to unite worlds. She felt humbled to participate.

The director went on to describe their Parliament's goal: developing responsible guidelines for quantitative research. Since applied quantum science could endanger the delicate cosmic spacetime fabric, it required a shared ethical approach.

Their first proposed quantum framework prohibited warp-speed travel and unfocused wormholes. Such technology recently came perilously close to shredding a neighbouring galaxy's chronal structure.

"We believe calibrated transit portals are vital for peaceful exchange between worlds," the director explained. "But unsafe applications could destabilize the entire local cluster."

Eva and Vikram joined a circle of diverse alien diplomats and began debating respectfully. Recognizing the threat posed by quantum distortions, they helped draft judicious policies embracing discovery while avoiding harm.

The Parliament sessions were rigorous but exhilarating. During recess, Eva and Vikram toured Pandora-3's labs with childlike awe. Their biotech, quantum computing, and even artistic disciplines operated at astonishing levels.

One evening, Eva gazed up at Andromeda's glittering expanse from a pinnacle overlooking the capital. Vikram joined her and followed her gaze skyward.

"Hard to believe there are civilizations out there even more advanced than the Watchers," Vikram remarked. "What mysteries do you think they've uncovered?"

Eva nodded. "We're like children playing in the tide pools compared to the ocean depths still unexplored. All we can do is keep wading deeper, one step at a time."

Their conversation was interrupted by gentle tremors vibrating at the pinnacle. Eva instinctively tensed before realizing the Watchers around her seemed unconcerned.

Vikram raised an eyebrow. "Is this a scheduled gravity wave calibration, or should we be worried?"

"Surface harmonization only," a Watcher reassured him. "We monitor all quantum fluxes closely."

But over the next several days, Eva noticed the gravitational vibrations growing more frequent and pronounced. She conferred with Watchers scientists, who seemed perplexed.

"The waveforms bear anomalies requiring further analysis," one told her. "Likely a natural cosmic event." But his expression remained troubled.

Late the next night, Eva was startled from sleep as violent quakes rocked the compound. Alarms blared wildly as Watchers rushed to monitoring stations.

One scientist studying the readout turned to Eva, his mental voice tinged with panic. "Our worst fears manifest. It appears a black hole has entered this galaxy, and its massive gravity well is rupturing cosmic spacetime!"

III

Part Three

10

Quantum Entanglement

Eva rushed with the Watchers to their advanced stellar cartography chamber. Holographic displays mapped the galaxy in exquisite detail, but she noticed ominous anomalies in the star formations surrounding Pandora-3.

Entire nebulae appeared distorted, and gravity wave amplitudes were spiking erratically. All evidence pointed to a rogue black hole careening through the region, wreaking havoc in its wake.

"How could your monitors have missed such a massive object until now?" Vikram asked one of the lead researchers, Zephyr.

"We suspect its quantum waveform origins," Zephyr replied. "Rather than a traditional singularity, this appears to be a knot of intensely entangled dark matter that spontaneously collapsed into a nascent black hole."

Eva's mind raced. Quantum-entangled particles exhibited instantaneous connections over any distance due to their linked wave functions. Somehow, clouds of entangled dark matter particles had coalesced into a tangled knot with gravitational force.

"Is there any way to predict this cosmic tie's movements?" she asked Zephyr. "Right now, it seems to be passing near Pandora-3. But if it wanders too close, your planet could get ripped apart."

Zephyr manipulated the holograms, charting the distortions along the black hole's path. "Now that we can detect its gravity shadow, I believe we can project its trajectory. Fortunately, our world appears safe for the moment."

He enlarged a model of the black hole itself—a writhing mass of infinite density distorting spacetime. "The greater risk is that it pulling other objects into its accretion disc. The resulting cataclysm could accelerate universal entropy."

Eva scrutinized the cosmic behemoth with frustration. After all her work helping the Watchers safeguard space-time, a random dark matter linkage threatened to undermine it all. There had to be a way to disentangle this knot before it grew beyond control.

In a flash of insight, she remembered delocalization, the quantum mechanical principle underlying entanglement. In essence, it breaks non-local bonds between particle pairs. Could matter caught in this black hole's grip likewise be freed by severing its quantum connection?

Eva proposed her theory to Zephyr and the researchers. They contemplated the radical notion solemnly.

"In principle, a precisely calibrated energy beam could delocalize the ensnared particles," Zephyr finally said. "However, the required power levels would be immense."

"What about harnessing vacuum energy flux?" suggested another Watcher. "We abandoned those methods long ago as perilous. But perhaps applied judiciously here, under strict controls, it could be worthwhile."

A tense silence followed as the team considered this dangerous option. Since renouncing risky quantum experiments, the Watchers have resisted utilizing such technology. But they may now have no choice.

Zephyr eventually turned to Eva, resolve settled on his alien features. "If we cannot disentangle this knot, both our civilizations will ultimately suffer. We ask for your help delocalizing the quantum entanglement threat."

Eva steeled herself and nodded. She would need to once again trust the Watchers' ability to constrain the chaotic vacuum forces they would be unleashing. The alternative—allowing the black hole to keep ingesting matter until no galaxies remained—was unthinkable.

They began assembling the components for a highly focused delocalization beam. Vikram provided engineering insights alongside the Watchers' physicists, while Eva tweaked the quantum field parameters, praying the calculations were correct. If she made any errors, the results could be apocalyptic.

Days later, the complex apparatus stood ready, aimed precisely for the heart of the writhing black mass on the holograms. Zephyr gave Eva the signal to activate. With a held breath, she initiated the delocalization sequence.

For endless moments, time seemed to stand still. Then finally, the holograms showed the black hole begin to untangle as ensnared particles were liberated by the dissolving quantum connection. Within minutes, only harmless dark matter clouded the screens.

The Watchers stood in reverent silence before Zephyr rested a slender hand on Eva's shoulder. "Once again, you have helped us conquer a cosmic trial. We are ever in your debt."

She shook her head. "No, the honour is mine. Our two people make a formidable team. No quantum challenge can withstand us."

11

Schrödinger's Cat

In the weeks after the rogue black hole incident, Eva and Vikram extended their stay on Pandora-3 at the Watchers' request. Their expertise with applied quantum mechanics proved invaluable for training the next generation of researchers.

Eva was also delighted to continue learning more about the fascinating alien culture. In their leisure time, she and Vikram often wandered the thriving capital city, marvelling at its graceful architecture and lush gardens.

One day, they came across an intriguing spherical building unlike the others, constructed from stone with a towering metallic spire. Curious, they approached and were greeted by an enthusiastic young Watcher.

"Welcome to our probability research institute," he conveyed telepathically. "New insights into cosmic uncertainty are discovered here daily."

As he led them inside, Eva studied conceptual holograms lining the curved walls. She recognized representations of classic quantum thought experiments like the double slit and Mach-Zehnder interferometer.

"These founding principles laid the groundwork for our advanced research," the Watcher explained. He gestured to scholars calibrating exotic equipment. "Their current focus is unravelling the dimensional intricacies of multiverse theory."

Eva was impressed by how eagerly the Watchers pursued knowledge, whether through abstract theory or practical invention. Their culture truly valued the scientific spirit.

As they completed the tour, one display in particular caught Eva's eye - the iconic framing of Schrödinger's cat. The paradoxical thought experiment had loomed large in humanity's exploration of quantum superposition and probability.

"Ah, I see you are intrigued by state vector duality," their guide remarked.

"On Earth, Schrödinger's cat is considered a seminal scientific puzzle," Vikram said.

The Watcher's large eyes grew thoughtful. "To our people, it represents how cosmic truths can be interpreted in different yet equally valid ways. The cat both exists and does not exist, until observed."

Pondering this perspective, Eva felt the human conceptualization of the experiment was somewhat limited. The Watchers perceived quantum realities as an infinite expanse.

12

Observer Effect

Eva spent the next week at the probability institute, fascinated by the Watchers' interpretation of quantum theory. By embracing paradox and uncertainty, their science achieved an intuitive elegance missing from human approaches.

One evening, while adjusting some research equipment, Eva noticed a faint shimmer pass through the lab. She looked around in confusion, but the Watchers seemed not to have noticed. She shrugged it off as an optical illusion caused by staring at display screens too long.

The next night, however, she clearly saw ghostly figures drifting near the lab walls before blinking out. Alarmed now, she rushed over but found only empty space.

"Did you see those wavering shapes?" she asked a nearby researcher named Zeta.

"Observing anomalous optical artefacts is common when manipulating probability fields," it replied calmly. "I would not be concerned."

But the bizarre visions kept appearing, lasting longer each night. Eva finally confided in Vikram, worried something was interfering with the institute's sensitive quantum experiments.

"Zeta said this phenomenon is normal, but I'm not convinced," she told him. "Whatever is causing this needs to be investigated before it affects research integrity."

Vikram agreed to help run diagnostics on the equipment after hours. That night, they initiated scanning protocols, but the tests found no system errors or malfunctions. Perplexed, they were about to quit when a vividly defined silhouette emerged near the very machine they were inspecting.

Eva leapt up to examine the area, passing her hand through the ghostly shape. "There's definitely an unknown energy signature here disrupting the quantum fields!"

"But where is it coming from?" Vikram wondered aloud. "And why can't the Watchers detect it?"

As they scrutinized the readings further, Eva grasped the cryptic source— their diagnostic scanning waves were somehow spawning the anomalies. The energy beams were interacting with the lab's probability fields to materialize transient phenomena.

"We have to stop the scans immediately," she told Vikram. "Our observations are collapsing superimposed quantum states into these phantom projections!"

He quickly terminated the equipment's active mode. As the waves died down, the mysterious shapes slowly faded back to nothingness.

"Of course—the observer effect," Vikram said. "By trying to analyse the anomalies, we ended up creating them!"

Eva shook her head ruefully. In their curiosity, she and Vikram had unconsciously demonstrated one of quantum theory's most fundamental principles: that the act of observation influences the system being observed.

"No wonder the Watchers appeared oblivious," she realized. "Their intuitive grasp of probability fields allows them to make passive observations only."

Chastened, Eva explained to the researchers that she and Vikram had accidentally induced the anomalies they saw. The Watchers, for their part, were amused by this distinctly human response to the unfamiliar.

"Your desire to understand is admirable," Zeta communicated. "But the cosmos reveals its mysteries only when we align our consciousness with its flow."

Taking this lesson to heart, Eva embraced the Watchers' approach when anomalies appeared in the future. Rather than actively probing, she practised mindful observation without expectation. Gradually, she learned to detect quantum distortions without collapsing their delicate superpositions through excessive measurement. By harmonizing with uncertainty, paradox flickered into clarity.

IV

Part Four

13

Wave Function Collapse

A few nights later, Eva was assisting some researchers in enhancing the probability lab's matter transmission technology. By precisely controlling quantum superposition, the Watchers could beam items over vast distances.

Eva was calibrating the receiving platform when a burst of garbled matter took form, coalescing into a misshapen blob. Cursing, she quickly shut the system down before more transportation errors could occur.

"Apologies, we are still smoothing out waveform overlaps," the lead researcher Omega informed her. "Without proper phase synchronization, superposed quantum states can collapse into unpredictable configurations."

Eva nodded, familiar with the phenomenon. According to quantum theory, subatomic particles existed as probabilistic waves until measured or observed. This caused the wave function to instantaneously "collapse" into a fixed state.

She helped Omega and his team adjust the wave emitters to better sustain superposition during transmission. By incrementally optimizing the harmonics, they successfully beamed a data crystal across the lab without distortion.

But as the testing continued, the calibration proved stubbornly inconsistent. Even minuscule ambient energy fluctuations caused abrupt wave function collapse. The research was halted out of frustration.

"These failures make no sense," Omega said. "The system showed flawless waveform integrity yesterday."

Eva bit her lip in thought. "What if an unknown factor is deliberately tampering with your quantum projections? Like a blind spot in your sensors."

Omega tilted his head curiously at her suggestion. He opened his mind to commune with his researchers, discussing Eva's theory.

"A rogue influence manipulating our experiments is conceivable," he finally conceded. "But it implies a cunning intellect. What being would attempt such deception?"

A sinister possibility took shape in Eva's mind: the Tau Ceti. The hostile alien race was known to employ stealth wormholes for deception. And they held a grudge against humans after losing the Pandora-3 wormhole war.

"We need to run a delta wave echo scan," she urged. "It will reveal any camouflaged objects in the quantum field."

The Watchers complied, activating the advanced scan. Almost immediately, the monitors displayed a faint shimmer, quickly forming into a small Tau Ceti scout ship flickering into view.

"An interloper!" Omega exclaimed. "How did it penetrate our perimeter?"

"They must have shadowed one of our ships back from Earth," Eva said grimly. "Then they used their cloaking tech to spy on your research."

But why? As if in response, the transmitter powered up, having been covertly hacked by Tau Ceti. A targeted matter stream beam erupted from the platform straight at Eva.

She leapt away just in time to avoid being struck by the defective transmission. The beam passed through empty space before hitting the wall, instantly spawning a writhing black mass that clung unnaturally.

"They weaponized the waveform collapse," Vikram realized. "Any target would dissolve into an unstable mess."

Omega quickly disabled the transmitter. "You have saved us from a grave threat. We are in your debt once more."

As Watchers defense ships surrounded the scout, Eva shook her head ruefully. Tau Ceti's devious sabotage had nearly brought disaster yet again. She should have guessed her old foes were behind the anomalies. Their vendetta clearly lingered, even with Earth and Pandora-3 now united.

This was a war the Tau Ceti could never win, Eva thought resolutely. The bonds between humanity and the Watchers would only grow stronger. They would face whatever came next together.

14

Many Worlds Interpretation

In the weeks following the Tau Ceti sabotage attempt, Eva doubled down on her collaboration with the Watchers to fortify their scientific defences. She knew her nefarious enemies would not give up easily.

While reviewing data from recent projects, a transmission from Earth caught Eva's eye. It was from an old colleague, Professor Nakamura, announcing a major discovery—evidence his team had detected parallel universes.

Eva swiftly contacted Vikram to discuss the news. "Nakamura was always obsessed with the many-worlds interpretation," she said. "But only in theory. Do you think his team actually found proof?"

Vikram looked thoughtful. "The math supporting parallel realities has been around for centuries. And lately, some quantum experiments have seen indications. If anyone could confirm it, Nakamura could."

Eva knew the respected physicist's reputation for rigour was unparalleled. "We have to try and replicate his results here on Pandora-3. Access to the Watchers' technology could allow us to take the first steps in communicating between universes!"

She arranged a conference with the Watchers' top astrophysics researchers to propose reproducing Nakamura's experiment. Although generally sceptical, they agreed it was worth investigating. The scientific trove such exploration could unlock was invaluable.

"We will provide our most precise quantum telescopes," their lead scientist, Delta, assured Eva. "Any cosmic vibrations between worlds will be detected."

It took weeks to calibrate the exotic dark matter detectors to scan for ultra-faint particles from a multiverse. Vikram wrote algorithms to analyse the subtle energy signatures Nakamura described. Late one night, the systems finally aligned.

"We're receiving echoes of what seem to be alternating universal states!" Vikram said it excitedly. "I think we're glimpsing parallel probability fields."

Eva stared at the cascading data in awe. Nakamura was right—the many-worlds interpretation was true. Diverging timelines spin off infinitely from each quantum interaction, like fractal branches.

As the Watchers expanded the scope of their scans, Eva noticed abnormal similarities between some universes' waveforms. Eerie resonances hummed between them.

"Those vibrational matches make no sense," she murmured. "It's as if parallel realities are being deliberately synchronized somehow."

A creeping unease filled her as she studied the anomalous data. There were too many conflicts amid the chaos. It reminded her of patterns found in engineered systems, like a fingerprint.

"Someone is coordinating those realities," Vikram voiced her suspicion. "But the technology required would be god-like."

A grim realization crystallized in Eva's mind. "Or Tau Ceti," she said coldly. "We know they utilize exotic spacetime manipulation. They could be exploiting parallel universes to bolster their power."

If so, this was a dangerous escalation in the aliens' capabilities. By aligning quantum probabilities across multiple worlds, they essentially controlled destiny. Eva shuddered, imagining her home reality moulded to serve Tau Ceti's unfathomable agenda.

She immediately contacted the Andromeda Council, warning of Tau Ceti's brazen violation of cosmic order. Stopping them would require collaboration between the greatest minds in the known galaxies.

If there were any constants between worlds, Eva knew it was the spirit of discovery and freedom she shared with the Watchers. No matter how many versions of reality the Tau Ceti chained together, somehow, they would find a way to set existence free again.

15

The Butterfly Effect

In the weeks after detecting the Tau Ceti's alarming interference across parallel worlds, Eva worked relentlessly analysing their meddling. She hoped to find a weakness that could be exploited before the damage was irreversible.

The data showed the aliens artfully modulating quantum probabilities through each universe they infiltrated. They were playing cosmic puppet masters, tweaking the most sensitive conditions to nudge events toward some sinister outcome.

But this godlike manipulation came with grave risks. The quantum wave functions in each world were intrinsically fragile; even tiny perturbations could get amplified over time. This was known as the "butterfly effect"—a flutter of wings triggering a distant hurricane.

Eva realized that, while advanced, the Tau Ceti were not omniscient. They failed to account for the inevitable turbulence in the systems they sought to control. With care, she could introduce countervailing fluctuations to disrupt their schemes.

When Eva detected a relatively isolated alternate world being influenced, she knew it was time to test her theory. After briefing the Andromeda Council, she and Vikram volunteered to intervene.

The Council was wary but accepted this trial run. They generated a wormhole connecting Eva's border facility to the coordinates of the endangered parallel Earth. She and Vikram steeled themselves and stepped into the shimmering portal.

Emerging on the other side, they found a city nearly identical to home but eerily devoid of life. Stagnant air carried a faint metallic trace. Vikram checked his scanner watch.

"Elevated sulphur particle levels - a by-product of wormholes," he noted. "The Tau Ceti must use this world as a gateway hub."

Eva's device mapped pockets of infrasound waves pulsating through the area—a sign of active quantum probability tampering. They were definitely in the right place. Then came the hard part.

Eva led Vikram to the tallest skyscraper overlooking the financial district, an area showing significant alien influence. She knew even tiny disruptions here could have an outsized butterfly effect.

Her scanner isolated three precise energy nodes embedded along the tower that served as quantum anchors for the Tau Ceti to manipulate this world's timeline.

Eva and Vikram got to work dislodging the anchors, using resonant pulses to disrupt the probability waveforms. It was a delicate process; too much interference could break the timeline entirely.

With the final anchor destabilized, Eva entered one last set of oscillations she believed would counteract the aliens' meddling. As the energy waves rippled out across the city, she held her breath.

Minutes later, her scanner showed the insidious infrasonic signals fading. Tau Ceti's hold on this reality was unravelling! Their delicate manipulations proved fragile in the face of the dynamic quantum forces they underestimated.

"One down, infinity to go," Vikram said with satisfaction as they prepared to return home.

Eva allowed herself a smile. Her audacious gambit had worked—a tiny flutter of wings to reverse engineer a distant hurricane. Though the battle was just beginning, with the infinite possibilities of the multiverse on their side, hope remained.

16

Causality Loop

When Eva and Vikram returned from their unauthorized intervention in the parallel timeline, they braced themselves for repercussions. But the Andromeda Council was amazed by their success in countering Tau Ceti's meddling.

"Your subtle methodology could be the breakthrough we need," Council leader Alpha told them. "With appropriate care, this approach may be repeated in other worlds."

Buoyed by this support, Eva and Vikram partnered with key allies among the Watchers and friendly galaxies to map a strategy. They targeted vulnerable alternate Earths being manipulated by the Tau Ceti through wormhole transit networks.

In each infiltrated world, their teams implemented surgical probability fluctuations much like Eva's initial test case. Fighting manipulation with manipulation was a dangerous game, but it proved potent for unravelling the Tau Ceti's invasion.

After liberating dozens of timelines, the allies detected seismic quantum energy bursts emanating near the aliens' home world. It appeared this guerilla resistance was destabilizing Tau Ceti's interdimensional infrastructure.

"Their whole operation is based on precise timeline control," Vikram speculated as they studied the readings. "Once chaos enters the system, cascading failures become inevitable."

QUANTUM WATCHERS: THE NEXUS PARADOX

But he and Eva knew the volatile aliens would not accept defeat lying down. When a wormhole link opened unexpectedly near her border facility, Eva sensed a confrontation was imminent.

Squads of Tau Ceti warships exited the vortex, unleashing weaponized warp bombs on the planet below. Watchers' defences quickly returned fire, but the initial attack left significant casualties.

Amid the carnage, one small scout ship broke through the planet's shield perimeter and touched down just outside Eva's base. She assembled her emergency team as a lone alien envoy disembarked and approached.

"You have interfered with our trans dimensional Dominion for the last time," the Tau Ceti leader warned via translator tech. "Your species will pay dearly for its insolence."

Eva stood her ground. "The universes you try to dominate do not belong to you," she said firmly. "Whatever you unleash, we will endure and rebuild."

The alien's eyes flashed with chilling ferocity. With a flick of his wrist, he opened a wormhole portal and disappeared back inside his ship. But strangely, the vessel did not attempt to leave.

Puzzled, Eva scanned the area, and her blood ran cold. The opened gateway had locked into a self-sustaining causality loop timeline - anything nearby would be trapped forever.

As the ship hovered menacingly over the camp, Vikram voiced the terrifying realization. "They're going to drag us into that wormhole loop as a punishment!"

Eva's mind raced for alternatives. They had mere minutes before the ship fired its chronal anchoring beam. Was this latest ruthless innovation of the Tau Ceti inescapable?

In desperation, she calculated a way to overload the loop's binding quantum field, snapping their base back into normal space-time flow. But the energy discharge would be enormous, putting her people at risk. This solution came with a heavy price.

Turning to her comrades with sorrow, Eva outlined the situation. "I can break us free, but the cascading forces may fracture this planet to its core. Your families would…"

Her voice faltered with emotion. But the Watchers responded immediately, volunteering for the mission that could end their civilization, to save uncounted more from the Tau Ceti's wrath.

With a heavy heart, Eva initiated the desperate overload sequence. As reality warped and bent around them, she could only pray their sacrifice would not be in vain. That somewhere in the multiverse, life would carry on - free from the monsters trying to enslave it.

17

The Grandfather Paradox

Eva opened her eyes, confused to find herself in what looked like a university science lab. Had the causality loop rupture sent her somewhere in the past?

She froze as a figure entered wearing vintage 20th-century clothes, then gasped in shock. "Granddad!?"

The man looked at her in confusion. "Excuse me, miss? I don't believe we've met."

Eva's mind reeled. According to old photos, this was her grandfather as a young quantum physics student at MIT in 1962. The overload must have thrown her back through time as well as space.

Hoping not to disrupt the timeline, Eva tried acting casual. "Sorry, you just remind me of someone I know. My name is Eva."

Her grandfather chuckled. "Well, pleased to meet you, Eva. I'm Isaac. Are you one of Professor Davidson's new research assistants?"

Eva nodded, going along with the assumption. She needed to learn more about where and when the wormhole had dropped her. Maybe there was a way back to her proper place in history.

As Isaac showed her around the retro lab, Eva's mind buzzed with the implications. Time travel theory held that interacting with one's own ancestors could trigger paradoxes altering history. The "grandfather paradox" scenario was famous—what if she accidentally caused Isaac's death now? She'd inadvertently erased her own future!

But, Eva reasoned, if she was careful not to disrupt the causal chain of events, perhaps she could safely observe this glimpse into her family's past. This chance meeting with her young grandfather was too poignant to avoid entirely.

Isaac proudly demonstrated cutting-edge equipment, like analogue mainframes and first-gen lasers. "One day, we'll use these tools to unlock the deepest mysteries of space and time," he mused.

Eva hid a smile. If only he knew his granddaughter would one day work alongside aliens to manipulate wormholes and quantum probability. The future was coming faster than he realized.

Their conversation was cut short when Isaac was called to help a colleague with an engineering dilemma. As he left, Eva's eyes landed on a folder of classified military schematics. Quantum teleportation theories... in the 1960s? Her curiosity was piqued.

Casually, she began photographing the research with her scanner watch, wondering if these forgotten designs could be useful in the future. She was so focused that she didn't notice Isaac returning until he spoke sharply.

"Eva, what are you doing? Those documents are restricted!"

She froze, realizing he had caught her stealing classified data. Thoughts racing, she tried explaining it away as curiosity, but the suspicious look on his face said he didn't buy it.

As Isaac went to summon security, Eva knew she had run out of time. She had to escape back to her present before she caused irreparable damage to this history. With a heavy heart, she activated her scanner's emergency wormhole protocol.

A shimmering portal opened, and with an apologetic last look at her grandfather, Eva stepped through. She emerged gasping with relief back at her base in the proper year, the lab folder still cradled in her arms.

Checking historical logs, she saw no anomalies had erupted from the encounter. By some miracle, she appeared to have avoided disrupting the timeline after all. The future was intact.

But seeing her grandfather so long ago had opened her eyes to the fragility of time itself. She knew she would need to tread carefully in whatever quantum journey came next.

18

Bootstrap Paradox

After her unsettling trip to the past, Eva tried focusing her quantum research closer to home. But analysing her grandfather's documents brought the challenges of time travel back to the forefront.

The classified schematics sketched technology eerily similar to current Watcher innovations: detailed quantum teleportation theories and prototype chronoportation devices. The anachronistic data made no sense.

Puzzled, Eva consulted Watcher engineers to trace the temporal origin point of key spacetime manipulation technologies. When the results came back, she stared slack-jawed in disbelief.

"These innovations shouldn't exist," she told Vikram. "The knowledge seemingly appeared from nowhere—no incremental development, just sudden existence! It violates causality."

Vikram scrutinized the data with unease. "A bootstrap paradox: ideas sent back in time from the future enabling their own later creation. But those causal loops could destabilize our entire timeline!"

Eva's mind reeled as the scope of the threat became clear. Knowledge from Watcher discoveries had clearly spilt backwards to 20th-century Earth via reckless time travellers.

This had seeded a dangerous temporal loop—present technology predicated on knowledge from itself in the future. And her scan of the classified documents had completed the loop, transmitting the schematics to her past self.

"We have to find where these leaks originated and prevent them," Eva insisted. "Any further contamination of the past risks our universe unravelling through paradox!"

Vikram nodded solemnly. As experts on quantum probability, they both knew the immense dangers posed by unregulated time travel and unchecked temporal manipulation. Action had to be taken before it was too late.

Working feverishly through nights and days, they developed chronic-encryption firewalls and causality mapping systems to contain anachronistic data. Each loophole patched brought some stability, yet the source of the rogue information remained elusive.

However, their sensors soon detected subtle wormhole aftershocks near Saturn—signs of unauthorized timeline interference. Confident they had found the paradox origin point, Eva and Vikram took a scout ship to investigate.

As the blue giant planet loomed large in their viewscreen, the readings grew ever more erratic. "There!" Vikram shouted, pointing toward an unfamiliar structure hovering just beyond the rings.

It appeared to be a drifting research station, heavily shielded to avoid detection. As they cautiously approached, a massive wormhole portal opened, swallowing the station before vanishing.

"The anomalies were coming from that base all along," Eva said grimly. "We have to track where it went before more damage gets done."

Returning to Pandora-3, they tinkered with chronoportation dynamics until locating the station's unique spacetime signature. Whoever they were, these renegades kept slipping away. But each jump left faint entangled particle traces, and the trail was growing fresher.

As a final wormhole opened, the quantum signature sent back the definitive identification: Tau Ceti! Eva and Vikram plunged into the vortex, ready to confront the perpetrators.

They emerged within the space station itself, weapons drawn, only to freeze in shock. The crew manning the systems was human! Rogue agents meddling with forces they scarcely understood.

Reluctantly, Vikram and Eva shut down the dangerous operation and began clean-up procedures to patch the ravaged timeline. The motives would come later. For now, protecting causality itself took priority.

As they escorted the defector team home for judgment, Eva shook her head in dismay. Whatever misguided goals drove this betrayal, the risks of paradox were too profound. Humanity would need to be vigilant and united on the quantum frontier.

V

Part Five

19

The Arrow of Time

Eva stepped through the shimmering portal onto the campus of the Galactic University, marvelling at the diversity of lifeforms around her. Beings from over two hundred worlds across the known galaxies convened here to explore the mysteries of cosmic time.

She was attending the University's symposium on temporal mechanics, her first since completing the advanced chronophysics program years ago. Since then, she has led daring expeditions probing the quantum depths, making discoveries that redefine humanity's concept of time itself.

Today she would share those radical findings with some of the most brilliant minds in the chronoverse. The implications were profound, predicting temporal vectors orthogonal to the established arrow of time. Her research pointed to modes of being not confined to cause and effect.

The presentations were held in a spectacular rotunda representing a scaled orrery of the Andromeda galaxy. As the sessions commenced, Eva was reminded of the many late nights spent here furiously debating wormhole theory and relativity with classmates from exotic worlds.

When her turn came, she took a deep breath and activated the holographic displays. "Esteemed researchers, we stand on the cusp of a new era," she began. "My expeditions into deep time suggest our universe contains dimensional folds hidden beyond perceived chronology."

Murmurs rippled through the auditorium as Eva described the quantum eddies and probability currents her ships had mapped while circumnavigating a black hole. Within these churning gyres, the subjective experience of time flowed erratically, sometimes even reversing.

"By analysing gravitational waveforms in the churning spacetime, I found regions where the arrow of time seems to curl backwards, while others diverge sideways. Our linear conception of temporality is only one facet of existence."

Eva's findings received a polarized response. Some attendees saw her research as recklessly speculative, while others were intrigued. In passionate discussions after, she engaged sceptics and allies alike.

"The discoveries you report could reshape civilization's entire world-view," an avian philosopher named K'Tass mused. "But we must take care when challenging orthodox chronology. Perilous distortions of causality could result."

Eva nodded thoughtfully. "You are right to urge caution. We still grasp so little about time's true, higher-dimensional nature. But I believe a unified theory will emerge in time, fusing these disparate arrows into a coherent whole."

Her long-time mentor, Professor Hyysa, clasped her shoulder warmly. "Once again, your work astounds, Eva. Know that there are many eager to advance these journeys of knowledge and imagination. Though the way forward is daunting, we walk it together."

Buoyed by this support, Eva returned home filled with renewed purpose. Hints of time's secret tapestry were coming into focus, guiding the way toward a future civilization unfettered by simple linearity. A society free to dance between modes of temporal experience at will.

She stood again under the flickering aurora borealis, echoing eons yet unloved. In those shimmering streams, Eva glimpsed the first traces of creation's next age, where souls might sail on possibilities unbound by epochs past and pending. Aching with hope, she set her compass to those distant shores.

20

Retro Causality

Eva stood before the Quantum Temporal Commission, still reeling from the revelations about the rogue human faction exploiting bootstrapped wormhole technology. Unchecked use of such advanced systems posed an existential threat to the integrity of the spacetime continuum.

"We are still assessing the damage done to our timeline," she told the assembled council members gravely. "Now that the leaks have been contained, we must ensure no causality violations remain."

HerTau Ceti counterpart, K'Rath, chimed in. "The foolishness of these renegades endangers more than just your people. If paradoxes shatter local space-time structures, our home worlds will suffer as well."

"Then we are decided," spoke wise Hyron of the Eldarii. "A quarantine shall be enacted until a proper Ministry of Time is established."

Eva knew this historic ban on unregulated time travel was necessary. And the Ministry could formulate the safe frameworks needed to resume controlled chrono-tech research. Still, looking out at her pioneering colleagues, she mourned the loss of momentum.

With the accord finalized, she returned home to a bittersweet celebration. Though proud of their accomplishments, the team now faced an uncertain future. Their very life's work was indefinitely suspended.

While programming the base's systems to begin automated shutdown, an alert suddenly flashed: unauthorized chronoporter activation in the research bay! Rushing to investigate, Eva was astonished to find Vikram powering up an experimental time engine prototype.

"Eva, you're just in time," he said, looking up with a smile. "I've solved the regulator issues and am ready for a test run."

"Vikram, you know we can't," she responded sadly. "The Council banned all chrono-travel."

"A short jump won't hurt," he insisted. "With your navigational skills, we can chart a smooth return curve before the timeline gets disrupted."

Staring into the shimmering portal Vikram had opened, Eva wavered. The chance to voyage through history once more was profoundly tempting. But violations now threatened more than just paradox…they betrayed the trust of their community and all who relied on a stable cosmos.

Mind settling into resolve, Eva deactivated the engine and placed a hand on Vikram's shoulder. "I want this as much as you," she said gently. "But we have to look beyond ourselves. The chronoverse will be waiting when we are ready."

Vikram lowered his eyes, conceding her wisdom. As they exited the hangar, Eva took one last longing glance at the dormant machine. She had to believe the Universe unfolding as it should, in its flawless singular order.

But over subsequent nights, Eva's dreams were haunted by visions of fantastic journeys through Aeons she would never see. Restless and conflicted, her thoughts kept returning to Vikram's tempting words - "one quick trip." In darker moments, she even found herself devising ways to bypass the sentinel chronobeacons monitoring the planet.

Two weeks after the ban, Eva was jolted awake by blaring proximity alarms. Rushing outside, she gasped to see the skies rippling with unfamiliar auroras— a sure sign of unauthorized wormhole travel nearby. Vikram stood staring up in dismay.

"Who could have…?" he began. But the truth sank in as they exchanged glances. Somehow, their future selves had returned to visit this forbidden past. The implications were chilling. Would they disregard all caution and unleash catastrophe? Or had their carelessness already bred some unforeseen disasters in epochs yet to come?

21

Quantum Tunneling

Eva could not tear her gaze from the unusual auroras flickering in the night sky—evidence that chrononauts from her future were brazenly flouting cosmic law. What urgent mission compelled their reckless arrival in this forbidden past? It seemed she and Vikram would someday violate every vow sworn to protect the timeline.

Moments later, an advanced small ship emerged from the atmospheric distortions and touched down smoothly nearby. Holding vigil over the dark valley, Eva and Vikram watched with bated breath as two silhouetted figures emerged. One tall, the other slender... Could those shadows be themselves?

"I know you have questions," the tall figure spoke gently as they approached. "But temporal integrity prevents us from explaining too much. Just know that events are moving too swiftly - we had no choice but to risk this intervention."

"Please," implored Eva, "at least tell us what actions to avoid that send us down this dangerous path."

The shadowy figures exchanged glances, then shook their heads sadly. "The crisis results from a discovery still ahead in your timeline," the woman said. "Forewarning would only accelerate its consequences."

With that cryptic warning, the duo turned to depart but suddenly froze - local chronal signals had unexpectedly shifted temporal phase! Without proper shielding, even a passive presence here threatened the visitors' native quantum waveform resonance.

Eva realized with dread that above-average solar neutrinos must have triggered the shift—an unpredictable hazard of any backward leap. Now a terrifying race was on to return the travellers to their origin moment before deadly quantum decoherence set in.

She and Vikram urgently transferred power to the visitors' chrono-drive while rapidly configuring an emergency particle accelerator to reinforce compatible quantum states. But the theory was one thing - could they successfully tunnel a pair of complex organisms through collapsing probabilistic barriers and thread the temporal needle home?

Power levels rising, the accelerator hummed to life, establishing a viable phase path into the future. But micro-distortions still riddled the projector beam. Eva despaired that their crude field modulation lacked the precision required to pass intact organisms.

Yet, observing the pair's fading waveforms, she noted curious patterns in the quantum noise - subtle yet familiar correlations she had seen in her grandfather's long-ago schematics. Of course! Adapting the accelerator with counter-phased tensor beams based on those lost designs produced a clean synchronic field.

As the ship powered up, Vikram activated the makeshift tunnel. For agonizing seconds the quantum barrier held firm. Then finally it yielded, the visitors' energy signatures resonating true as they slipped into the probabilistic manifold. A moment later, the ship vanished into its home epoch.

Turning to Vikram in exhausted relief, Eva knew desperate questions still loomed. What was the nature of the coming discovery to set their fates so dangerously adrift? Had their aid today saved the future or only enabled it?

But she sensed also a deeper truth - past and future were bound as one. Whatever paths unfold, she and Vikram would face them together, through courage and sacrifice if called for. Such was the essence of spacetime itself.

As violet quantum auroras faded from the cosmos, calm returned to the valley. Tomorrow's horizons lay once more veiled, with both darkness and wonders in store when light rose anew. But for now, in this moment, there was peace.

22

Superposition

In the weeks following the surreal visitation from her future self, Eva found her focus constantly drifting as she struggled to resume normal research duties. But how could she ignore the cryptic warnings that awaited down her timeline? What discovery was imminent that could compel such reckless time travel?

During a refuelling stop at a remote space station, Eva decided to confide in the wise Centauri captain K'var, whose lengthy lifetime granted her a unique perspective.

"Respected K'var, if you knew events were coming that would corrupt all you value, how far would you go to prevent them?" she asked.

K'var studied her with large emerald eyes before responding. "To have seen one's future is both a gift and a burden," he said slowly. "We may wish to steer fate in another direction, yet often our actions instead lock destiny upon its course."

Seeing Eva's frustration, he continued gently. "Consider the photon—here and there at once until observed. The time likewise remains potential until it is lived. You dwell now in the superposition between the shadowed past and the veiled future. Make the most of this open present."

His words lingered with Eva as she travelled home. Perhaps anticipating mysteries ahead blinded her to gifts already in hand. She resolved to lead her team with renewed passion in the day that shone before them.

Shortly after, their outpost received an urgent relay from the central home world. Scientists analysing a distant gas giant had detected inexplicable energy signatures—evidence of an unknown civilization harnessing primal quantum forces once considered impossible to tame.

"We believe a sapient species dwells within the planet itself, manipulating gravitational potentials through applied superposition principles," the lead researcher Zeta elaborated. "But this power, if mishandled, could warp gravitational fields on a cosmic scale."

Eva exchanged an uneasy look with Vikram. Quantum gravity experimentation of such magnitude had been theoretically banned for ages due to extreme volatility. But it seemed these enigmatic aliens—code-named "Striders"—either ignored or were unaware of the prohibition.

"How do we make contact without provoking conflict?" Vikram asked with a frown. "Any direct intervention risks triggering the very gravitational collapse we hope to avoid."

Eva bit her lip in thought. Vikram was right—a misstep here could breed a galaxy-wide cataclysm. She considered the photon paradox again, existing everywhere and nowhere until pinned by observation to a single state.

"We have to remain in quantum superposition—observing without collapsing probabilities through interference," she proposed. "Deploy stealth reconnaissance drones to gather intelligence. Perhaps understanding the Striders' motives can prevent disaster."

The Council approved Eva's plan. She trained tireless AI drones in non-invasive approach patterns before scattering them silently into the gas giant's churning atmosphere. Over months, the probes mapped the planet's interior, transmitting footage back through entangled quantum channels.

A vivid picture began taking shape: the Striders were bipedal methane-breathers who inhabited a layer of the planet where water existed as exotic ice. They were scientifically advanced but culturally isolationist, valuing knowledge above all else.

"They're manipulating quantum-scale gravity waves through concentrated tunnelling reactions," Vikram realized. "Individually minute, but the combined superposition effect produces staggering energies."

Indeed, the Striders seemed unaware that their esoteric experiments were beginning to resonate with the planet's core, threatening seismic fractures. But how to warn them without hostile retaliation remained vexing.

After careful deliberation, Eva dispatched a single probe bearing gifts—data crystals containing lessons on harnessing gravitational potentials safely. She knew the odds of changing their path were slim, but she had to try. Once contact was made, events would unfold as they would. All that mattered was having chosen light over darkness.

VI

Part Six

23

Quantum Teleportation

Eva stepped onto the teleportation platform, preparing to demonstrate the new instant matter transmitter that could revolutionize planetary commerce and travel. She nodded to Vikram at the controls; years of research had led to this moment.

"Initializing quantum waveform projection," announced Vikram. "Encoding molecular scan parameters across entangled pairs."

Eva felt a tingle as the platform hummed to life, calibrating to match her quantum state. A Chat could now instantaneously transmit her exact atomic configuration to a receiver pad far away, where she would reassemble.

"All systems are ready," Vikram said. "On your mark for teleport test in 3...2...1..."

In a flash, Eva's surroundings vanished. Oddly, she felt no sense of motion during the transit. In the next instant, she smoothly rematerialized in the receiver booth. Success!

The teams cheered, but Eva hushed them impatiently. "Send the lab mouse through now," she urged Vikram. Until a living specimen teleported intact, the system remained unproven.

Vikram placed the mouse in the transmitter booth and initialized the sequence. Moments later, they held their collective breath as the mouse took form on the receiver pad, blinking up curiously. The quantum teleporter was operational!

As Vikram ran diagnostics to extend the range, Eva mentally reviewed the technology's core principles. Matter transmission was achieved by directly converting particles into pure quantum information—the waveform—which was instantaneously projected to any destination with an entangled receiver.

Eva had studied the quantum foundations, while Vikram engineered the systems to translate macro-objects between waveform and particle states. Their combined expertise enabled teleportation from fantasy to reality.

"Excellent work," their Director Zeta commended them when the milestone was officially confirmed. "With further refinement, this technology will enable instant cargo delivery and travel across the world. A new era dawns thanks to your brilliance."

Potential applications whirled through Eva's mind. Teleportation could bridge worlds and galaxy clusters in the blink of an eye. But troubling implications also loomed. She insisted exhaustive security protocols be implemented before public use. If hijacked, such a tool could enable unspeakable horrors.

Despite precautions, an unauthorized teleportation occurred weeks later. Investigating the incident, Eva was shocked to discover the perpetrators were militant Tau Ceti—bitter remnants clinging to obsolete notions of conquest. Somehow they had spied on Eva's research and built their own system to raid a lunar archive.

"Thankfully, no one was harmed this time," Vikram said. "But this proves the technology's dangers in the wrong hands. We should discontinue further development before it's too late."

Eva refused, arguing the benefits were too great if handled responsibly. "Discovery itself is neutral; only intentions create benefit or harm," she maintained.

After thwarting multiple other attempts at misuse, anti-teleportation protests escalated beyond the fringe. Public anxiety mounted that such power could never be contained. Some have declared it inherently amoral.

"Will you reconsider?" Vikram asked. "Continuing risks normalizing something so potent it could unravel social order."

"You underestimate people's adaptiveness," Eva countered. "This is a tipping point, yes, but progress must march onward."

Before the debate concluded, a shocking mass demonstration destroyed the main research teleporter, plunging the future into uncertainty. Only time would tell whether society moved past fear toward uplift or if the destruction of technology avoided an existential threat. In quantum mechanics, both outcomes had probability.

24

Quantum Computing

Eva stepped into the secret underground facility, eager to see the quantum computer prototype Vikram had told her about. Known only to their research team, this experimental system aimed to harness the infinite processing potential of qubits—quantum bits—to solve computational challenges once thought insurmountable.

"Thanks for coming, Eva," Vikram said, leading her down a gleaming passage. "If successful, this could be one of the most significant technological leaps since the classical silicon age."

They entered a chamber where an intricate lattice of shimmering metal and glass components formed an enclosed structure the size of a room. Inside the meticulously calibrated shell, an array of charged particles was suspended in a quantum superposition of states.

"Each qubit exists as a one, a zero, or both simultaneously," Vikram explained. "Their indeterminate state allows for massively parallel computation. We've packed 500 qubits so far, pushing the envelope of quantum coherence time."

Eva nodded, transfixed by the device where fundamental forces of nature had been precision-engineered into a digital synthesizer.

"So what will this quantum computer actually do?" she asked.

"One application is chemical simulations," Vikram said. "The variable parameter space for molecular interactions explodes exponentially. This machine can model physical processes far beyond classical computers."

He activated the system, initiating the laser pulses and cryogenic controls that allowed the delicate quantum field to stabilize. Eva watched eagerly as the qubits were manipulated through calculations.

"For an initial test, we had it analyse a simple peptide chain. The simulation was more than a billion times faster than our existing supercomputer."

Eva shook her head in wonder. By embracing the fuzziness permitted by quantum physics, they had effectively achieved computational power magnitudes beyond traditional binary logic.

Over the next few weeks, staggering insights emerged as the system rapidly unlocked complex organic reactions, protein folding configurations, and cellular mechanisms. Eva could scarcely keep up with researching the real-world applications being uncovered.

However, an ethical dilemma arose when the military proposed deploying the computer's predictive algorithms for bioweapon development. While profoundly opposed, Eva knew refusing the request outright could result in the project being seized under classified status.

"The technology itself has no allegiance; only its users do," Vikram said when Eva sought his counsel. "Should we accept some moral sacrifice if it means maintaining oversight?"

Eva sighed, disheartened to think their visionary work might be turned toward destruction.

In the end, she submitted a deliberately sabotaged bioweapon model, knowing the flaws hidden within would render any resulting agent ineffective. The military moved their funding elsewhere, believing quantum computing did not live up to its promise.

Relieved, Eva prayed future generations might use such machines solely for humanistic advances, never oppression. She did not know what tomorrow would bring, but today at least, science's light still shone as a beacon of hope against the darkness.

25

Quantum Cryptography

Eva hurried to meet Vikram, weaving between the bustling shops and cafes lining the capital's tree-shaded promenade. When she arrived at their favourite noodle bar, he was already seated with a concerned expression.

"Sorry I'm late; I got caught in tram delays," she said, sliding into the booth. "So what was so urgent?"

Vikram hesitated, then spoke quietly. "Have you heard whispers of the Chronos Protocol? Rumours are swirling about their shadow research into quantum cryptography and communication security."

Eva nodded furtively. The secrecy around this government initiative gave her chills. In the name of national security, who knew what moral lines they might cross?

"Some of our old academy peers joined the Protocol out of naivete," Vikram continued. "But I'm hearing reports of illegal technology developments. They seem to be weaponizing quantum mechanics into tools of surveillance and oppression."

Eva gripped her glass tightly, mind racing. Theoretically, quantum computing could crack any existing encryption. And manipulations of quantum entanglement allowed for perfectly secure information exchange. Such knowledge in the wrong hands could have terrifying consequences.

"We have to do something, Vikram," she urged. "If Chronos Protocol goes unchecked, no communication will be safe from their prying eyes. Our society will become a panopticon!"

"I agree, but we must be cautious," he advised. "Chronos likely has moles throughout the government to protect their interests."

They left the cafe alone to avoid drawing suspicion. That night, Eva contacted trusted allies from the cybernetics institute to discuss countering Chronos' dangerous advances.

Their most pressing goal was to develop quantum encryption robust enough to resist Chronos' codebreaking supercomputers. After weeks of running simulations, they created encryption keys harnessing quantum indeterminacy—secure codes that even infinite processing could not crack.

The next phase was preventing Chronos from intercepting communications through quantum tapping of the photon transmissions. For this, Eva and her team exploited quantum entanglement to share one-time pads that were detectable if illicitly observed.

As these quantum cryptography systems were covertly deployed, Chronos' spying efforts hit a wall. Baffled by the uncrackable codes and secure channels, they scrambled to develop countermeasures.

Eventually, a network of anonymous whistleblowers began exposing shocking details of the Protocol's constitutional violations. As public outrage mounted, politicians were forced to condemn the program they once rubber-stamped, fearing losing their seats.

"Chronos' days are numbered," Vikram told Eva as more revelations emerged. "You dealt them a crushing blow without stooping to their level. Well done."

But a key figure in the protocol still remained—an elusive operative known only as Omega. This mastermind was determined to regain the upper hand for Chronos.

One evening, Eva was shocked when Omega himself contacted her on a secured channel. "Your interference has only prolonged the inevitable," he stated cryptically. "Chronos exists to safeguard civilization. Progress demands sacrifice."

"Some lines should never be crossed, no matter the justification," Eva responded firmly.

The shadowy agent laughed coldly before terminating the transmission. Eva knew this was not the last she would hear from him.

Weeks later, her network detected massive energy spikes at a remote Chronos research base. She realized with dread that it could only mean Omega had successfully activated the Protocol's rumoured ultimate doomsday device—a weapon harnessing the quantum vacuum forces at the very heart of spacetime itself.

With the stolen technology now at his fingertips, the twisted visionary called again. "Today we shape destiny," Omega proclaimed. "After this demonstration of our new power, your resistance will end. A new era begins for our world!"

26

Spooky Action at a Distance

Eva waited anxiously in the Chronos Facility control room, along with the research teams tapped to validate the radical new power system Omega had unveiled. Harnessing quantum vacuum energy has long eluded the greatest minds, but somehow this shadowy group has achieved it.

Around her, Chronos engineers made final adjustments to ominous equipment shrouded under tarps. Eva exchanged an uneasy glance with Vikram; they both had grave misgivings about this technology in the hands of zealots. But any objections were silenced as the reclusive Omega entered to excited murmurs.

"Welcome all," he greeted the assembled crowd. "You are about to witness history." With a dramatic flourish, he drew back the tarps, unveiling a sleek reactor-like device encircled by elaborate amplification rings.

"Behold our quantum vacuum capacitor!" Omega proclaimed proudly. "Within this chamber, the very quantum foam of spacetime itself is harvested through precision entanglement modulation."

Murmured exclamations rippled through the control room. Most believed vacuum energy extraction was unattainable. Yet here was proof the protocol had done the impossible.

Omega continued, "Once activated, negatively charged virtual particles drawn in are separated from their anti-particle pairs before they recombine. Their captured virtual photons will power a new world!"

He signalled the start-up sequence. The capacitor rings began undulating as the core chamber glowed fiercely brighter. Eva watched the monitors nervously as energy figures climbed, praying runaway chain reactions could be contained.

"Initiating transfer beam," Omega called out. At his prompt, the capacitor discharged, focusing a brilliant lance of light into an adjacent engine structure. The beam seemed to draw in surrounding photons as it connected.

The engine activated with a resounding roar, its output gauges redlining at previously unimaginable levels. The Chronos engineers cheered at their success. The era of limitless energy had begun.

But during the celebration, Eva noticed concerning readings - the vacuum device was drawing in ambient heat at an exponential rate as its core temperature plunged toward zero kelvin.

"Something's wrong," she told Vikram. "It's acting like a runaway heat vortex - once begun, it may never stop!"

Before Vikram could respond, proximity alarms blared. The dense mass accumulating around the core had destabilized local space-time. The entire facility was at risk of getting crushed into a singularity!

As technicians rushed to disengage the system, Omega whipped out an ionizer weapon, his eyes crazed. "No one shuts this down! The capacitor is only beginning to show its potential!"

He fired wildly around the control room, but Vikram tackled him, allowing security to finally wrest away the gun and detain the unravelling architect of catastrophe.

With Omega subdued, Eva urgently interfaced with the control systems, fighting to reverse the critical energy build-up as local gravity went haywire. To her relief, an emergency override finally managed to shut the device down, stabilizing the distortions as power levels normalized.

In the aftermath, facility technicians soberly dismantled the compromised technology. Omega and his extremists were removed from Chronos leadership, and their visions of power were discredited through unhinged science.

Eva knew research into quantum vacuum energy would one day proceed responsibly, but only once culture had evolved ethically to match such a leap. For now, slow and steady progress was the wise path forward.

She squeezed Vikram's hand affectionately. Once again, their intuition and bravery had prevented disaster. What new trials lay ahead remained uncertain, but she had faith that together they would meet each other with compassion and conscience as guiding stars.

27

Quantum Zeno Effect

Eva hurried excitedly through the research complex, eager to see the new stasis field generator Vikram wanted to demonstrate. After years of studying quasi-particle containment, his team believed they had developed a working prototype. If successful, the implications would be profound, enabling the indefinite suspension of quantum systems.

Arriving at the experiment chamber, Eva embraced Vikram warmly. "I came as soon as I got your message! It sounds like you've achieved a real breakthrough."

Vikram smiled, his eyes twinkling. "We're on the cusp of something big here. The stasis generator exploits wave function control to freeze quantum states in temporal limbo. Matter caught in the projection field is instantly put into a state of quantum suspension!"

He led Eva inside, where a control booth looked out on a test platform. The stasis generator itself was an imposing toroidal frame surrounding the platform, with multiple emitters dotting its rim.

"Its operation relies on selective quantum measurement - the Zeno effect," Vikram explained. "Frequent observation of particles inhibits state change. Under constant measurement, the waveform essentially becomes static!"

Eva nodded thoughtfully. It was an ingenious application of the quantum principles they had spent careers exploring. By continually tracking particle spin with focused scans from the emitters, change was instantly detected and counteracted, locking systems into temporal stasis.

"Shall we see it in action?" Vikram said, powering up the generator. The emitters began pulsing in complex patterns as the space above the platform shimmered in a projected quantum field.

Vikram placed an organic sample into the field. As the emitters stimulated the Zeno effect, the sample's atoms frozen in place, unaffected by passing time. Eva watched the readout in awe. A biological specimen in flawless quantum suspension!

Over the following weeks, Eva worked closely with Vikram's team to refine the stasis generator's capabilities. Larger and more complex objects were successfully immobilized for longer durations. The potential applications were astounding: perishable goods could now be preserved indefinitely, and biological specimens could be safely secured in temporal isolation.

However, during an ambitious test, an unexpected power surge overloaded the projection field. The quantum waveform protecting the temporarily suspended matter instantly collapsed.

Eva gasped in horror as the organic sample rematerialized as a misshapen, mutated mass, as though centuries of entropy rushed back at once. Though safely disposed of, the grim sight shook her deeply.

"This technology is too unpredictable and dangerous in its current form," she told Vikram gravely. "Until the risks are better understood, further research must halt."

Vikram reluctantly agreed. For all its potential, the accidents demonstrated they had yet to truly grasp the nuances of quantum chronal manipulation. They dismantled the prototype generator.

"One day, with wisdom guiding our hands, humanity will revisit this frontier," Vikram said wistfully. "Not as conquerors of time's flow, but as students, ever awed by its unfolding mystery."

Eva smiled and gripped his shoulder. Once again, their conscience prevailed over ambition. What future wonders or perils awaited down untrodden paths, only time would tell. For now, today was enough.

28

Measurement Problem

Eva walked briskly to the university hall, eager to hear Dr. Amira's progress on her groundbreaking experiments addressing the quantum measurement problem. Solving this physics enigma could reshape their understanding of reality itself.

Inside the packed auditorium, Dr. Amira took the podium amid hushed anticipation. "Colleagues," she began, "today I will share extraordinary new evidence suggesting consciousness alone causes wave function collapse in quantum systems."

Murmurs rippled through the hall. The role of consciousness in quantum mechanics was hotly debated. Mainstream theory supported passive observation, but Dr. Amira was proposing active influence.

Her research had centred around rigorous double-slit studies with matter-wave diffraction. In a vacuumed chamber, individual particles were projected through dual slits, then measured at a capture plate. When unobserved, interference patterns suggested the particles passed through both slits as spread waves. But introducing any measurement made the pattern collapse into particles passing through one slit only.

Dr. Amira replicated this but added an unconventional variable - human observation. Volunteers meditated during the particle projection phase, directly observing the process with expanded consciousness without any physical recording. The results were astonishing—interference still gradually collapsed.

"Consistent across all trials, the act of conscious human observation alone influenced the wave function physicality," Dr. Amira revealed. "Mental focus alone determines concrete reality."

The presentation was met with uproar as academics interrogated her radical conclusions. Vikram turned to Eva nervously. "This research could reshape global physics. But does observing really influence reality so profoundly?"

Eva nodded thoughtfully. "If consciousness alone collapses the wave function, then the mind creates reality. But contemplating existence as wholly subjective raises deep questions. We must discuss this with Amira further."

At the symposium reception, Eva and Vikram drew Dr. Amira aside. "Your work promises groundbreaking revelations," Eva said earnestly. "But as colleagues, we are concerned about larger societal impacts. How can we advance together responsibly?"

Dr. Amira sighed but nodded. "You are right to be cautious. I admit in my enthusiasm, I did not fully consider dangerous misinterpretations." She reflected for a moment. "Objective studies should continue confirming my results. But philosophical implications require measured thought and debate."

Relieved, she welcomed constructive counsel, and they eagerly discussed safeguards against dogmatism, keeping science open and self-critical. The trio committed to leading through ethics as much as discovery.

In the following months, Dr. Amira's experiments continued under formal peer review, validating her stunning conclusions. Meanwhile, Eva and Vikram liaised with global forums on the metaphysical dimensions.

The measurement problem had unlocked the next revelation in humanity's awakening. But greater truths required even broader perspectives. Life's essential oneness would be the beacon lighting the uncertain way ahead.

Eyes fixed on that distant shore, Eva meditated each dawn on her place within creation's infinite beauty. She observed not to constrain the world's wild potentials, but to leave them unbound by limitations of thought. Wave and particle, dreamer and dreamed - all were welcome in this spacious awareness.

29

Double Slit Experiment

Eva tried to contain her excitement as the university quantum physics club prepared to recreate the iconic double-slit experiment that had captivated her since freshman studies. While a common educational demonstration, witnessing wave-particle duality first-hand was exhilarating.

The experimental setup looked simple - a quantum particle emitter directed at a solid plate pierced by two parallel slits, with a detector screen behind to capture the light pattern. But the mysteries it unveiled were profound.

"Remember, light is both particle and wave," Vikram explained to the gathered students. "As particles, photons pass through one slit or the other. But as waves, they diffract through both slits simultaneously. This creates an interference pattern on the detector from the overlapping light waves."

He activated the emitter, projecting individual photons toward the double slits. Eva watched eagerly as the detector screen revealed the predicted interference bands. The probabilistic cloud of waves and particles persisted in quantum superposition right up until the moment of detection.

"Now, let's observe which slit each photon passes through," Vikram continued. "This should collapse their wave functions, eliminating the interference."

He engaged precision photon detectors at each slit. As expected, the interference pattern instantly vanished, leaving only two distinct particle clusters corresponding to each slit pathway. Actively measuring one state precluded the simultaneous measurement of another.

One student raised her hand. "So the photons choose either particle or wave nature the moment we observe them?"

Vikram smiled. "In a way, but human observation doesn't fundamentally change their behaviour. The act of measurement interacts with the system enough to force a definite state."

Eva added, "It's less about what nature 'chooses' and more about interaction limiting its possibilities into just one outcome."

She never tired of pondering the philosophical implications. How many simultaneous versions of reality overlapped until awareness crystallized into just one option?

After the students dispersed, Vikram came over to Eva. "Thanks for your insight today. The double-slit experiment remains profound even after all this time. It reveals how consciousness and reality intertwine."

Eva nodded. "Observing alters what we observe. To truly see light's full potential, we must learn to watch gently with an open mind."

Their conversation was interrupted by an automated security alert. Unauthorized access was detected in restricted laboratory sectors. Vikram and Eva exchanged alarmed glances—those labs contained dangerous experimental technology.

Rushing to investigate, they discovered a break-in by unknown intruders who had raided classified quantum research specs. Security footage showed a thief replicating the university's double-slit apparatus, clearly intending to misuse the principles demonstrated.

"We have to find those thieves before they unleash something hazardous," Vikram said worriedly. "Who knows what havoc they could wreak by exploiting wave-particle duality?"

Eva's mind raced with possibilities, each more ominous. By selectively collapsing or preserving quantum superpositions, one could manipulate countless systems to achieve desired outcomes. In malign hands, the power could be terrible indeed.

Working feverishly with university security, they tracked energy readings from the stolen equipment to a warehouse on the city outskirts. There they found a makeshift quantum optics lab where the thieves were already employing the apparatus for avaricious gain.

By bombarding financial networks with photons placed in simultaneous states, they altered market system behaviour at opportune moments. An ingenious scheme powered by quantum advantage.

"Halt at once before you tear the very fabric of reality!" Eva demanded.

Spooked, the thieves attempted to destroy their makeshift lab but were stopped in time by university authorities. The dangerous stolen equipment was reclaimed before irreversible damage was done.

"Another reminder that knowledge itself has no allegiance," Vikram mused. "The double slit mystery reveals reality's flexibility, but we must honour that gift with wisdom."

Eva nodded, contemplating the lesson as they watched the criminals hauled away. The quantum realm was an amphitheatre of infinite potential - and sentient beings its performers, co-creating each new act.

30

Quantum Nonlocality

Eva could hardly contain her excitement as the Andromeda Exploratory Vessel dropped out of faster-than-light travel, bringing into view the spectacular cosmic phenomenon she had spent years studying: - The Boötes Supernova.

This ancient exploded star was a laboratory for testing quantum non-locality—the mysterious ability of entangled particles to instantly influence each other at vast distances. Proving that quantum effects operate outside normal spacetime could reshape their understanding of reality itself.

"Preparing to launch the entangled meson probes," announced Vikram, calibrating the ship's powerful payload accelerators. "We've got one shot to align this perfectly across light years."

Eva checked the navigational vectors. "Set triangulated coordinates to those three pulsars. Their gravity wells should slingshot the probes into intersecting beams across the supernova's leading shockwave front."

Vikram nodded and initiated the launch sequence. Twin streams of accelerated subatomic particles, intricately entangled, shot toward Boötes' massive debris field. Passing stars bent the beams across several light years until they were perfectly aligned.

Eva watched the displays eagerly as the probes finally intersected, unleashing a burst of energy bright enough to observe from Earth. "Now we see if manipulating one beam affects the other," she said breathlessly.

As Vikram oscillated one stream's quantum spin, the reactive modulation of the second beam answered the question immediately. The action was induced instantly across light-years of separation!

"By the stars, it works!" Vikram exclaimed. "Faster-than-light communication between entangled particles is real!"

Eva beamed, delighted that their audacious experiment proved non-locality beyond doubt. This discovery opened the door to instantaneous transmissions across vast distances, finally transcending the speed limit of light.

However, a cryptic quantum fluctuation then caught Eva's eye. "Look, the excess energy at the intersection point isn't dissipating normally..." she said worriedly.

Vikram studied the readings and blanched. "You're right, it's cascading into a runaway entanglement chain reaction!"

Alarms blared as the quantum discharge began spreading exponentially through the tenuous gas, sparking reactions with erratic effects. Charged particles blinked in and out of existence. Pockets of gravity were inverted randomly.

"It's brewing a quantum storm!" Vikram yelled above the din. "We have to escape before we get trapped here!"

He desperately tried manoeuvring the ship away from the growing spacetime anomaly. But the spreading entanglement field hit the hull, locking systems in quantum flux. Controls no longer responded, and sensors reported hull integrity failing.

Through the viewport, they watched in rising panic as a churning wormhole vortex began opening all around them, threatening to swallow the ship.

At the last moment, Eva had a desperate inspiration: intentionally overloading the still-active entangled probes to intentionally collapse their wave function coherence.

"Disentangling" them shut down the runaway reaction cascading from their linkage. Instantly the anomaly dissipated, safely reverting local space to normal stability.

As the vortex vanished, Vikram stabilized systems and piloted them to a safe distance. He let out a shaky exhale. "Incredibly quick thinking! You turned their entanglement against itself."

Eva nodded solemnly. "We witnessed the double-edged potential of non-locality today. But with care and conscience, we will master this new horizon."

No discovery was without risk, she knew. What mattered was cultivating the wisdom to wield new knowledge for flourishing, never destruction. As the stars called them onward, Eva embraced uncertainty as the lifeblood of existence - and their shared voyage just beginning.

31

Quantum Decoherence

Eva hurried down the university hallway, weaving between students, to reach Dr. Amit's quantum physics lecture, already in progress. She was eager to hear the esteemed professor's perspective on the spacetime anomalies at the leading edge of cosmic research.

Slipping quietly into a seat beside Vikram, she opened her holo pad to take notes as Dr. Amit highlighted unsettling data gathered from deep field scans. Strange boundary layers between galaxies showed increasing quantum decoherence effects—the tendency of matter to destabilize from coherent superposition into classical states.

"Something appears to be disrupting the quantum probability matrix underlying physical reality in those regions," Dr. Amit noted with concern. "Local space-time itself is losing definition and becoming fluid."

Eva shared an uneasy glance with Vikram. Could such dissolution of cosmic coherence explain the phenomenon of entire ships simply vanishing? Space agencies had recorded several such disappearances near decoherence zones, despite no equipment failures.

"So what could be causing this breakdown in quantum phase integrity?" a student asked.

Dr. Amit stroked his beard thoughtfully. "Exotic matter with inverted vacuum energy is a possibility. Such 'phantom' particles radiate negative pressure that can tear the quantum vacuum."

He pulled up an expansive galactic hologram. "If significant quantities of phantom matter accumulated in intergalactic mediums, the resulting decoherence could eventually reach catastrophic levels."

Eva felt a chill, imagining streams of inverted particles steadily eroding the quantum scaffolding of the universe like corrosive radiation. Matter and energy, as they knew it, would lose all meaning in the spreading void zones.

After class, she shared an idea with Vikram. "We should use the university's new zero-point field laboratory to test decoupling techniques for slowing decoherence. By modulating vacuum field harassment, we might provide a buffer."

Vikram agreed that it was worth investigating. In the strange matter lab, they configured force-field projectors to simulate quantum flipping effects, then bombarded the isolated zone with stabilizing photons. To their excitement, the rift-like fractures were reduced, successfully shielding a sample of normal matter.

"With some scaling up, these units could provide temporary decoherence abatement in affected space sectors," Vikram said enthusiastically.

But his excitement soon faded along with the grant funding for the unconventional research. The scientific consensus held that decoherence zones posed no immediate harm, just an interesting observational anomaly. And the wider world had concerns more pressing than such remote cosmological threats.

Frustrated but not deterred, Eva and Vikram continued simulations in secret, eventually producing a compact decoherence mitigation drone. They floated the proposal to concerned spacer guilds, who endorsed deploying a network of the drones to secure trade routes and early warning.

Setting up a makeshift mission control in Vikram's garage, they piloted the first drones into a fresh decoherence zone manifesting near a pulsar. As the drones interfaced with the quantum turbulence, spacetime distortions visibly receded, buying vital hours for ships to navigate clear.

When an endangered guild convoy was saved from certain doom, public perception finally shifted. Now heralded as visionaries, Eva and Vikram secured government funding to deploy drone fleets wherever decoherence bloomed.

Yet the expanding crisis weighed on them. At best, the drones delayed the inevitable, like fingers plugging a dike. And the decoherence waves emanated inexorably outward, wave after wave...

"There must be a source," Eva whispered, studying projections of decay sneaking closer toward inhabited worlds. "Some origin we can intercede at."

Vikram took her hand, equal measures of despair and hope in his eyes. "Whatever is eroding the quantum foam, we will find it. And together with wisdom, we will make this darkness retreat."

Though realistic about the challenges ahead, Eva embraced Vikram's resolute optimism. Mysterious and perilous the void may be, but life would kindle fires of laughter and love until the last star winked out. Heart and mind united could illuminate even the deepest unknown.

VII

Part Seven

32

Parallel Universes

Eva could hardly contain her excitement as the exploratory vessel prepared to activate its reality drive and make the unprecedented leap into a parallel universe. Years ago, the theoretical discovery of the multiverse sparked heated controversy across human civilization. But today, Eva and Vikram would be the first to gaze upon one of those alternate realities with their own eyes.

"Initiating reality shift sequence," announced Vikram, making adjustments at the control pillar. "We've locked onto the Delta-4 quantum signature from those satellites you planted. Setting course across the dimensional barrier now."

Outside the viewport, a shimmering rift began to open, spilling prismatic light within the void. Eva watched in awe as the iridescent colours merged into a passageway between realities. Vikram piloted them into the widening vortex.

Moments later, the ship emerged into a familiar star scape—yet subtly different. "Did we make it?" Eva wondered aloud.

"Long-range scans detect the magnetic variance," Vikram said, visibly excited. "We're definitely in Delta-4!"

As they flew toward the nearest solar system, Eva could see this universe diverging from theirs sometime in the 19th century. Planet-wide empires still vied for colonial power, and sailing ships filled space lanes.

"Incredible…" she breathed. "An alternate path, split off from our history, is now alive before us. All those pivotal moments that didn't happen"

Approaching Delta-4's version of Earth, they cloaked the ship to avoid disturbing the unsuspecting human society below. Through viewports, Eva saw ornate floating cities controlled by aristocratic houses, as ordinary citizens toiled on ravaged landscapes.

"They may look like us, but their world evolved down darker roads," Vikram said gravely.

"With compassion, perhaps we can guide them toward more just possibilities," Eva replied.

They spent the next few weeks covertly learning about this Earth. Eva's heart broke seeing oppressed worker castes crushed under the bootheel of industrial barons. She knew intervention was risky, but her conscience called her to action.

One night, she infiltrated a secure data archive, decrypted classified files on the ruling elite's abuses of power, and transmitted the revelations across global networks. As scandals erupted, angry uprisings finally toppled the corrupt regimes.

Returning to their ship, Eva saw Vikram's eyes shine with pride. "You gave them a chance to choose freedom. Whatever comes next is theirs to write."

Their mission was complete, and they prepared to return home with invaluable data on alternate human development. But as the reality drive powered up, proximity alarms suddenly blared a warning.

Confused, Eva checked external scans. "Massive gravitational shear waves are emanating from the portal…but why?"

"Its quantum aperture must be destabilizing!" Vikram said. "We have to shut it down before it ruptures both universes!"

He desperately tried emergency systems to deactivate the unstable wormhole. But the quakes intensified, fracturing the space around them. It was too late - the rending of realities had already begun.

Eva hugged Vikram close as a blinding light engulfed the ship. No words needed to be spoken to convey what was in their hearts. Though fate may separate them, their bond was eternal.

Eva found herself adrift in a formless void, stripped of all identity and memory. She panicked for a terrifying moment before realizing Vikram floated there too. In this infinite gap between dimensions, only love persisted.

Joining hands, they surrendered to the churning, indifferent cosmos reshaping itself anew. The multiverse's endless dance continued, realities winking in and out of existence with each cosmic breath. Between the stars, they shone on.

33

Quantum Immortality

Eva took a steady breath as she prepared to step into the quantum immortality scanner. This experimental device promised a journey beyond death itself using quantum superposition principles. By copying her entangled consciousness the moment before likely death, Eva could experience continuing life in a cloned body.

"Remember, this is only a trial run," cautioned Vikram, making adjustments on the control terminal. "We're testing the cognitive transfer process, not aiming for full corporeal reconstruction."

Eva nodded, giving Vikram's hand a grateful squeeze before entering the spherical scanner chamber. She settled into the activation pod, placing the psycho-conductive filaments onto her temples.

"Ready when you are," she said, attempting to calm her nervous excitement. After years of developing this technology, today could prove the concept viable.

"Initiating psychic waveform replication sequence," Vikram announced.

The pod hummed around Eva as flashes of light stimulated the synchronized firing of her neurons. She could feel her consciousness expanding, perceiving reality on both macro and quantum levels.

"Entering the critical phase," warned Vikram. "Temporary systemic paralysis will prevent actual cell death."

Eva's breathing halted as the pod induced a sudden convulsive seizure throughout her nervous system. As everything went dark, she felt her psyche detach, suspended mid-transfer in the probability matrix.

Moments later, Eva gasped as awareness flooded back. She was alive! Vikram rushed over, helping her out of the pod with an amazed expression.

"It worked!" he said joyfully. "Your disembodied consciousness was successfully encoded into the quantum probability field before material reconstruction."

Eva marvelled at the implications. By preventing true death, the scanner enabled a form of eternity, allowing users to persist indefinitely. With further research, true immortality was within reach.

In the following weeks, repeated tests consolidated the transfer process. Eva's consciousness could be reliably preserved as a self-contained psychic waveform when separated from her physical form.

However, an unexpected discovery gave them both pause. The transfer process was producing quantum duplicates—identical versions of Eva waking up separately after each trial run.

"It must be an unintentional bifurcation of your consciousness waveform," Vikram theorized with concern. "We're going to have to suspend testing until I can improve the confinement stream."

Eva nodded uneasily. The proliferation of "hers" was deeply unsettling, each feeling equally like the original. She insisted the duplicates be kept in limited sensory isolation to avoid psychological distress.

Over time, Eva noticed Vikram growing strangely distant and obsessive. One night, she discovered the awful truth: he had secretly continued developing the scanner technology and produced a fully-formed clone of her by extracting DNA from a hair sample.

When confronted, Vikram pleaded that his love for her drove him to perfect the technology faster—to never lose her. Horrified by his delusion, Eva demanded he halt the dangerous research immediately.

"But why do you get to be the 'original' Eva?" He countered angrily. "I could activate your consciousness in any of the inert bodies."

Eva recoiled from this warped manifestation of their dream. Immortality through technology had blinded Vikram to life's true meaning—the singular preciousness of each moment, never to be replicated.

She tearfully demolished the scanner herself, freeing the trapped psychic essences. Together, she and Vikram held a memorial for all the "Eva's" they had condemned to soulless limbo in pursuit of eternity.

From the ashes of misguided intention, Eva emerged wiser. Physical death awaited everyone, and no quantum trickery could undo its holy mystery. Life's singular flashes were gifts to be savoured and then passed on.

She would cease grasping so desperately and instead, walk gently into the unknown. What awaited beyond she did not yet know. But she trusted it held secrets grander than immortality.

34

Quantum Suicide

Eva stepped solemnly into the chronometry chamber, exchanging a stoic nod with Vikram at the control terminal. After exhaustive research, today they would finally activate the device that promised to answer the most fundamental question in quantum mechanics: was consciousness bound to a single reality or did it persist across a multiverse?

Known as quantum suicide, the experimental process relied on quantum immortality principles. The chronometer would initiate a sequence with only one survivable outcome, triggering a divergent reality split. By experiencing survival, Eva's consciousness would confirm existence simultaneously across probabilities.

"Commencing the 100 micro-iteration countdown," announced Vikram, initiating the chronometer's intricate systems. Eva settled into the focus pod at the core and took a deep breath. In a few moments, they would know if the many-worlds interpretation was truth or fanciful theory.

"Scanning neuronal topology" called out Vikram. "Entangling brainwave function with probability matrices..."

Eva's consciousness perceived dual realities flickering as the pod interfaced her mind with the quantum field powering the device. In one realm, the chronometer malfunctioned, discharging lethal dark energy tendrils. On the other, it shut down safely.

"Activating quantum bifurcation," said Vikram. Dizzying vertigo seized

Eva's mind as causality split. In one branch, screams of agony before darkness. In another, gasping relief as the pod powered down.

But only awareness in the living branch could resonate. By being alive to witness the result, Eva immediately confirmed that her consciousness persisted across the branched realities. Just as theorized!

Emerging joyfully from the pod, Eva rushed to hug Vikram. "It works—we exist simultaneously in alternate realities!" she exclaimed. "Consciousness remains continuous so long as at least one version carries on."

They celebrated the monumental confirmation of the many-worlds theory late into the night. With this device, Eva could directly experience the quantum multiverse and its infinite possibilities first-hand. Her mind reeled envisioning the revelations it could unlock into the deepest riddles of existence.

In the weeks that followed, repeated experiments expanded Eva's awe at life's endless forking paths. As her research notes filled with profound insights, she began to perceive reality itself through an enlightened lens.

However, the repeated near-death experiences soon produced concerning changes. Eva grew withdrawn, often staring right through Vikram when he spoke to her. She dismissed his worry that chronic quantum jumps were taking an emotional toll.

One evening, as they prepared to commence another bifurcation, Eva hesitated before entering the pod. "Each trial confirms my suspicion," she said distantly. "We are bound to no single history. Therefore, nothing done here has a purpose, as all possibilities play out."

Vikram's face creased with worry. "Your nihilism is the chronometer's side effect, corrupting thought," he implored. "We can't lose sight of science's humane application."

But Eva seemed resolved. "One reality is as fleeting as any other," she replied before suddenly sealing the pod and engaging the lethal sequence.

"No!" Vikram screamed in anguish as lethal energy discharged inside the pod, obscuring Eva's form. Falling to his knees, he dismantled the deadly device through heartbroken tears. In another realm, Eva lived on. But here, a light had gone out in all possible worlds.

35

Quantum Archaeology

Eva could hardly contain her excitement as the university's time-dilation shuttle prepared to voyage into the distant past. As an archaeologist, the prospect of directly observing ancient civilizations first-hand was a dream come true. By exploiting quantum gravity properties, the shuttle could traverse the centuries with ease.

"Temporal course charted for the Mesopotamian dynasty of Akkad, circa 2350 BCE," announced Vikram, the shuttle's chrononautics engineer. He gave Eva an encouraging smile. "Ready to take the deep dive into history?"

"I've been waiting my whole career for this," Eva replied eagerly. She settled into the passenger bay designed to buffer the temporal shift's radical forces. Through the viewport, a shimmering wormhole vortex began to open.

Moments later, the swirling colours resolved into a dusty plain surrounded by ancient stone architecture. Eva could hardly believe her eyes. "There the fabled ziggurat of Akkad!" she exclaimed. "We made it!"

Donning disguises from the period, she and Vikram cautiously ventured out to observe daily life inconspicuously. As Eva documented details with her concealed scanner, her heart swelled at being immersed in a world only known from buried fragments. The past was no longer an abstraction but a living, breathing reality.

When the risk of disrupting the native timeline grew too severe, they retreated back to the shuttle. "One day side-by-side was worth a hundred years of studying artefacts," Eva said breathlessly.

Vikram grinned. "I'd say our maiden voyage was a success. Where to next? The Library of Alexandria perhaps?"

As the shuttle voyaged into eras of wonder, Eva's appreciation for the vast tapestry of human history deepened exponentially. But an unsettling question loomed. If the past could be accessed, could it also be changed?

Her concerns seemed confirmed when the shuttle detected residual quantum turbulence swirling around the moment Vesuvius erupted as if an outside force had recently manipulated that event's probability wave.

"It looks like we aren't the only quantum archaeologists," Vikram noted worriedly.

Their scans eventually revealed a cunning corporation exploiting time travel to ransack history for monetary gain. By raiding lost treasures or strategically altering past events, they reaped massive profits in the modern era.

"These criminals are destroying the integrity of our shared heritage!" Eva said it angrily. "We have to shut them down before they do irreparable damage."

Devising a plan, Eva and Vikram intercepted the corporation's agents during an illicit jump to Pompeii. Battling them for control of their shuttle, Vikram and Eva lead the crooks on a reckless chase straight into a collapsing wormhole. Their ship was lost to the ages.

With the timeline marauders stopped, Eva and Vikram sealed their own technology away until culture had gained the wisdom to use it ethically. "Quantum archaeology has boundless lessons to teach," Eva reflected. "But our reach must not exceed society's grasp."

She returned to traditional archaeology with a renewed sense of awe at history's grandeur. As long as humanity kept reverently uncovering its past, there were always new depths of wonder yet to be revealed.

36

Quantum Archaeology Paradox

Eva stepped cautiously out of the university's time shuttle, still disoriented from the abrupt chronic-jump. Vikram sat at the controls, looking pale, trying to stabilize their damaged systems. Around them, the jungle thrummed with strange bird calls.

"Did we make it?" Eva asked breathlessly. "Is this truly Earth, 65 million years ago?"

Vikram nodded, awe breaking through the anxiety on his face. "We're smack in the Cretaceous period according to telemetry. The time storm's random wormhole must have flung us clear out of the modern era!"

Eva peered anxiously through the dense prehistoric foliage. They had been on a routine excursion observing the Renaissance when violent temporal distortions struck, sending them careening down the timeline.

She knew it would take days to recalibrate the shuttle for the immense return jump. Until then, they were trapped - but also presented with an unimaginable opportunity for archaeological study.

"We should take cautious samples of flora and fauna before leaving," she told Vikram eagerly. "Imagine the discoveries hidden in this ancient biosphere!"

He hesitated, doubt creeping into his expression. "Is that wise? We risk drastically impacting the evolutionary timeline."

Eva's shoulders slumped. Vikram was right—preserving cosmic integrity had to take priority over fact-finding. As scientists, their first duty was doing no harm.

Dejected, she watched through the shuttle port as Vikram worked to chart them an ultra-precise return path, accounting for Earth's orbital shifts. How she wished to tread where no human had before! To unveil life as it was, unvarnished by the eons. But virtue demanded restraint.

Hours passed uneasily. Then, just before Vikram completed the calculations, proximity alarms blared a warning. Something had breached the clearing surrounding the shuttle.

Eva froze in terror at the sight of a towering T-Rex trampling toward them. "Vikram, hurry!" she yelled. "If we don't jump to the future now, we'll be eaten for sure!"

But fate offered no such salvation. Vikram cried out as the tyrannosaur collided headfirst with the craft, sending it rolling sideways into the muck. Escape was now impossible; they were stranded!

As the T-Rex thrashed, trying to bite at the shuttle's hull, Eva knew they had only one option left, however reluctantly.

"Prepare the emergency sonic cannon," she said grimly. "Let's aim to scare it off without permanent harm."

Vikram quickly configured the non-lethal sound weapon and blasted the piercing frequencies at the prehistoric giant. It reared back with a pained roar before retreating into the dense jungle. They were safe—for the moment.

Settling back inside, Eva placed a consoling hand on Vikram's shoulder. "You were right to urge caution before," she said gently. "Our presence here already carries unknowable consequences. Our duty now is minimal interference."

Vikram sighed but nodded resolutely. Together, they would find a way to uphold the sanctity of the past, so that life's tapestry could weave itself forward as intended.

Their voyage home, when it came at last, would bear invaluable lessons about resisting the arrogance that time was theirs to exploit. To tread the chronoverse was to walk humbly between eras, stewards of cosmic memory more than its masters. Wherever the tangled quantum streams carried them next, Eva and Vikram now saw more clearly the ethics shining the way.

37

Nexus Point

Eva stepped through the shimmering wormhole onto the rocky surface of Volatilis-4, an unassuming moon orbiting a gas giant on the frontier of the Venturi Expanse. At first glance, nothing suggested this sphere's pivotal importance. Yet celestial scans clearly showed powerful and inexplicable quantum energies emanating from below its surface.

"This must be the nexus point linking multiple cosmic time streams," she said, turning to Vikram as he emerged beside her. "If so, manipulating its novel temporal mechanics could have reality-altering effects across local space."

Vikram nodded. "Our theories suggest this quantum crater anchors a convergence in probability lines from a multitude of quantum histories."

He pointed to unusual plasma ribbons arcing across the stars. "Even that celestial aurora seems aligned with the emanations. We've stumbled upon something profound."

Their research consortium had dispatched Eva and Vikram to investigate anomalies hinting at unique spacetime properties in this obscure corner of the universe. Though cautious of unknown forces potent enough to warp quantum causality itself, they could not resist this unprecedented discovery.

Moving carefully between the jagged rocks lining the crater's rim, they began assembling delicate instrumentation to analyse the eerie stellar radiance rising from below. Once sufficient data on the nexus point's nature was gathered, they could model its complex chronal mechanics.

"If we could consciously harness these wellspring energies, history itself could be reshaped," Vikram posited as he processed the exotic feedback.

Eva nodded thoughtfully. "In theory, yes. But tinkering blindly with cosmic forces of this scale would be reckless beyond measure. We must learn their deeper purpose first."

For weeks, they gradually mapped the nexus point's role as an axis where probable histories diverged, entangled, and birthed new realities. Eva was reminded of ancient notions of a World Tree or cosmic loom—metaphors for existence's underlying oneness.

But one evening, their research was interrupted by the shriek of engines. From the stars emerged an imposing warship, descending toward the nexus point with weapons primed. Through the gloom, Eva spotted familiar markings—the Tau Ceti Imperium!

"They must have detected the nexus too and seek to control it," Vikram realized with dismay. "With that power, no civilized world would be safe from their tyranny."

Eva's mind raced. The nexus point was too significant to allow its exploitation, even if protecting it meant their lives. Setting the station to self-destruct, they gathered the minimal data and samples possible before hastily opening an escape wormhole.

Just as they dove in, Eva took one last look at the sparkling crater as the first enemy salvos erupted. What wonders and horrors had that place witnessed across eons of cosmic history? For now, its secrets would remain veiled.

Back on their home world, Vikram and Eva's findings generated much debate. Some scientists urged harnessing the nexus point's potential if it could serve the greater good. Others counselled strict avoidance, arguing some powers were beyond ethical control.

In the end, the expedition's data was sealed away until society developed wisdom to match such dangerous knowledge. Whatever potential paths the quantum Nexus could enable, Eva maintained hope that compassion would light the way forward.

On calm nights pondering the universe's mysteries, her thoughts drifted back to that tiny, unassuming moon guarding its Well of Worlds. All Moments were one, after all. The future's shining web remained open, waiting for gentle hands to guide the shuttle of fate.

38

Time Crystals

Eva hurried down the university hallway, weaving between students, to reach Dr. Amit's quantum physics lecture, already in progress. She was eager to hear the esteemed professor's perspective on time crystals, an exotic new form of matter that subverted the normal flow of time within its quantum structure.

Slipping quietly into a seat beside Vikram, she opened her holopad to take notes as Dr. Amit highlighted unsettling data gathered from the new time crystals. Their atoms exhibited periodic motion without energy inputs, essentially generating perpetual motion—long thought impossible.

"By trapping their quantum particles in recurring temporal loops, these crystals represent a breakthrough in chrono-engineering," Dr. Amit noted. "However, scaling up such unstable, time-bending matter could have reality-distorting effects."

Eva shared an uneasy glance with Vikram. She knew radical research groups were already investigating weapon applications using matter caught in temporal recursion. Could disrupting time crystals' delicate quantum flows on a larger scale lead to a local time-jamming effect?

"So how do we responsibly advance time crystal research given the risks involved?" a student asked.

Dr. Amit stroked his beard thoughtfully. "Expanded study is warranted, but under controlled isolated conditions. Strict oversight and use principles must be established proactively."

After class, Eva met up with Vikram in the quantum computing lab. They found their colleagues abuzz with news—the lab had just synthesized its first tiny time crystal overnight. But the delicate process and precise tuning required made scaling the feat a distant goal.

Still, holding the glimmering millimetre-sized crystal, knowing its atoms spun outside normal spacetime, filled Eva with awe. "We have to take things slow with this one," she told Vikram seriously. "Who knows what havoc larger versions could wreak if misused?"

Vikram nodded. "Agreed. Until time crystals are better understood, we limit all production to the bare minimum for study."

In the following weeks, however, accidents plagued the lab. A compromised stasis field leaked irregular temporal waves, briefly entrapping three researchers in disjointed microseconds before the anomaly could be sealed.

Eva realized with dread that corporate spies must be attempting to steal their crystal synthesis methods. "We need heightened security measures before this escalates," she urged Vikram.

But their warnings went unheeded as bureaucratic obstacles stonewalled proposals for improved safeguards. Forced to take matters into their own hands, Vikram and Eva destroyed all existing time crystals and wiped the sensitive data.

"That should deter the cronies of corrupt profiteers," Vikram said. "Some doors are best left unopened."

Soon after, a law was passed banning time crystal development without a permit. While disappointed at the loss of knowledge, Eva was relieved that reckless minds would no longer run amok. The Universe ticked on whether or not eyes watched its clock.

Patience and care were the wise keys in chrono-engineering. In time, perhaps their civilization would mature enough to again work wonders with temporal matter. But for now, putting the time crystals to rest honoured cosmic rhythms grander than any human ambition.

When future generations looked back, Eva hoped they would remember pioneers who had not rushed heedlessly after breakthroughs but listened first to the resonance of creation unfolding as it should—no moment too fast or slow.

39

Chronology Protection Conjecture

Eva hurried excitedly through the university campus, weaving between students toward Dr. Roy's advanced seminar on temporal mechanics. As a pioneer in quantum chronodynamics research, Dr. Roy's insights into guarding spacetime integrity were profound.

Arriving at the auditorium, Eva slid into a seat beside Vikram and activated her hole-tablet, eager to record Dr. Roy's wisdom. The white-bearded physicist cleared his throat and began.

"Class, today we will discuss the intricate discipline of chronology protection," he said, conjuring intricate 4D models in mid-air. "This emerging science aims to shield causal continuity from disruption by unchecked time travel or other tachyon-based technologies."

Eva nodded along attentively, having witnessed first-hand the dangers of unchecked spacetime manipulation during experimental voyages. She still shuddered, remembering the sight of possible futures wrinkling away into the quantum aether like tissue paper in a fire.

Dr. Roy went on, "It is my firm belief that cosmic physics inherently resists paradoxes that could undermine existence. While closed timeline loops are permitted, excessive distortions trigger self-censoring processes - like feedback preventing an uncontrolled PA system from devolving into sonic chaos."

To demonstrate, he simulated a hypothetical time machine splintering its past, which then rapidly self-corrected by gravitationally repelling the disruptive technology. "Thus the 'chronology protection conjecture'—spacetime preserves its structural integrity."

After class, Eva met up with Vikram to grab coffee and discuss. "Do you buy Dr. Roy's theory that paradox is intrinsically impossible?" she asked.

Vikram looked thoughtful as he sipped his espresso. "I could see elegant self-correcting principles woven into the universe's quantum code. But full proof remains lacking."

Their conversation was interrupted by an ominous space-time turbulence alert. Rushing outside, they were stunned to see violent purple thunderheads boiling ominously over the campus.

"Chronal vortices are opening nearby!" Vikram exclaimed, shielding his face from the roaring temporal winds. "We have to take shelter before it shreds causal coherence!"

Sprinting through the building debris, they took refuge in the sub-basement as howling vortices shattered the space-time continuum itself outside. Watching in awe and terror as possible pasts, presents, and futures collided chaotically, Eva truly appreciated Dr. Roy's life's work safeguarding cosmic chronology.

When the cataclysmic storm finally passed, the university emerged miraculously intact, a testament to the universe's ability to absorb disruptions. Life resiliently persisted in whatever form it could.

As she and Vikram helped with rebuilding efforts, Eva's respect grew for the forces guiding existence gently through its most turbulent eras. The river of time always harmonized its rhythms, whether or not eyes watched its clock.

She knew future investigations would achieve wondrous and dangerous new heights. But come what may, somehow, the shining sea of stars endures. Of this, she held hope, however many tomorrows away.

40

The End of Paradox

Eva stepped cautiously onto the rocky surface of Volatilis-4, returning to the remote moon that had long obsessed her imagination. Years ago, she and Vikram had discovered this inconspicuous orbital body hid immense power, anchoring a nexus point connecting cosmic time streams.

They had fled before the zealous Tau Ceti could seize control. But now, Eva has led an authorized coalition to uncover the nexus's deepest secrets, hoping to harness its quantum energies responsibly to heal ruptures across the multiverse.

Her scans showed the familiar shimmering vortex churning within the moon's central crater. But this time, no hostile warships emerged. With care and wisdom, she knew they could illuminate the mysteries of creation itself here.

Assembling advanced instruments around the sparkling vortex, the research team began carefully probing its churning chronal strands. Eva monitored the incoming data in awe as the nexus's size and complexity rapidly exceeded all projections.

"Its probability matrix contains nearly infinite alternate time streams in entwined superposition," breathed Vikram, her veteran research partner. "We're looking at a core anchor for the quantum wave function of reality itself!"

Eva nodded solemnly. "Let's focus initial mapping on our native timeline.

Perhaps Nexus wave packets could repair recently destabilized eras."

Her people had overcome many perils, but flaws persisted across generations. Eva held onto hope that the Nexus could help heal old wounds that continued reverberating discordantly.

But manipulating even the tiniest fraction of the nexus's power proved extraordinarily difficult and dangerous. Like operating on a beating heart, the most precise touch risked cascading trauma. Despite endless care, each simulation Eva ran to protect her timeline's integrity seemed to only fracture it further.

Late one restless night on the lunar surface, Eva stared up at the harsh beauty of infinite stars, feeling despair encroach. Had she doomed her world by returning here? Were some powers truly beyond any mortal hand's ability to guide?

Vikram gave her shoulder a supportive squeeze. "Your intentions are peaceful, but remember, there are infinite ways for events to unfold. Sometimes love means holding space for life's own flowing."

His wise counsel turned her desperate grasping into open curiosity. They had uncovered this nexus with reverence; its purpose remained beautifully veiled.

A new understanding blossomed—there was no tool to crudely remake existence. This cosmic loom held space for empires to rise, civilizations to fall, and souls to wander and return, in its own timing. Eva's trying had obscured its perfection.

She opened her heart again to life's uncertainty and fragility. However frayed its threads grew, the tapestry remained sacred. Mending tears was wise, but only with care not to unravel greater patterns.

With Vikram's help, Eva erased their instruments' dangerous meddling and restored the Nexus Point's pristine state. They had approached a holy of holies. Now was the time for listening, not shaping.

Before leaving Volatilis-4, Eva stood once more on its craggy surface, feeling possibilities resonate from the quantum crater-like music. Existence danced on, paradox-free in its myriad expressions. All moments were one.

When future ages arrive more enlightened, the Loom of Worlds will be waiting. But for now, this moon's weaving was its own - endless, edgeless, and undreamt-of. Not hers to know, only witness. And that was miracle enough.